OLIVER TRACTORS

A CHIEF ENGINEER'S ACCOUNT~1940–1970

T. Herbert Morrell

Octane Press Softcover edition, August 2020
Text originally published as *Oliver Farm Tractors*
by Motorbooks, September 1997 under ISBN 978-0760303566
Octane Press, Hardcover edition, September 2012

ISBN 978-1-64234-027-3

Book design by Tom Heffron
Edited by Blanche S. Morrell and Judy Lumb
Copy Edited by Joseph Holschuh
Proofread by Tobias Gros

octanepress.com

Octane Press is based in Austin, Texas

Printed in the United States

DEDICATION

I dedicate this book to my wife, Blanche S. Morrell, who has been very supportive during my entire professional career. She has assisted in the editing of this book, provided many good suggestions, and helped me in numerous other ways.

T. Herbert Morrell
July 1996

CONTENTS

ACKNOWLEDGMENTS 7

FOREWORD 9

PREFACE 11

AUTHOR BIOGRAPHY 15

CHAPTER ONE
Company History 17

CHAPTER TWO
Development of the Fleetline Tractors 59

CHAPTER THREE
XO-121 Research Program 111

CHAPTER FOUR
The Super Tractors 121

CHAPTER FIVE
Another New Line: 1600, 1750, 1800, and 1900 Tractors 141

CHAPTER SIX
Special Tractors 157

REFERENCES 185

APPENDIX A
Tractor Models 187

APPENDIX B
Safety Standards 193

APPENDIX C
Development of Oliver's New Gasoline Engine 201

ACKNOWLEDGMENTS

I wish to thank those who have helped me with this book. Members of the Hart-Parr Oliver Collectors Association encouraged me to write this book and provided much information. Bob Tallman and Sherry Schaefer have been extremely helpful. Special thanks go to the Oliver and Hart-Parr collectors who permitted photos of their tractors to be included in the book.

Many people provided photos and historical information, including Leland (Skip) Hartwell, Kurt Aumann, and Mary Ann Townsend of the Floyd County Historical Society in Charles City. Keith L. (Punch) Pfundstein sent information and photos of Charles F. Kettering and his participation in the XO-121 research project. Glenn B. Bazen provided photographs of Lull Engineering uses of Oliver tractors. Doug Strawser provided information on the number of tractors and combines sold to Russia in the 1930s, and Edna Ladd gave some insight into Oliver's circumstances at that time. Samuel W. White Jr. has been interested in the book from the beginning and graciously agreed to write the Foreword.

I am grateful to Sherry Schaefer, Doug Strawser, Guy Fay, and Lorry Dunning for reviewing the final manuscript and, especially, to our editor, Lee Klancher, who was enthusiastic about the project and guided us every step of the way.

Because of her desktop publishing experience our daughter, Judy Lumb, was very helpful and spent many hours refining and editing. My wife, Blanche, our son, Dennis Morrell, and daughter, Becky Schmitz, have been very supportive and encouraged me to write this book.

T. Herbert Morrell
Owatonna, Minnesota
June 1996

FOREWORD

I am honored to be asked to write a foreword for this most remarkable and detailed book on the accomplishments of the Oliver Corp. Especially noteworthy are the author's personal contributions to the progress of Oliver's work and its recognition among the major U.S. and foreign tractor manufacturers.

Tractors are the backbone of a full-line agricultural machinery manufacturer. The farmer (and later industrial) customer, the sales and service personnel in the Oliver dealer network, the loyal and dedicated Oliver employees in the plants, branches, and overseas all freely expressed their admiration for the high quality, innovative features, productivity, and long life of our Oliver tractors.

My career with Oliver started at the South Bend, Indiana, plant before I finished graduate school and resumed after World War II. I served as Oliver president from 1960 to 1970 after White Motor Corp. purchased the Oliver Corp. in 1960. From my vantage point as chief executive officer in the Chicago office and my constant travels to visit our branches, dealers, and competitors, I witnessed first-hand the high regard accorded Oliver's outstanding products and the special praise reserved for the Charles City plant engineers and their plant personnel. What a joy it was to note the ready acceptance of the many forward-looking firsts and competitive innovations that Herb Morrell and his staff brought to our Oliver Corp.

I enjoyed a most pleasant personal and professional working relationship with Herb Morrell. I supported his programs, congratulated his dedicated staff, heralded his designs, and took the lead in seeking recognition for Oliver products—their features, productivity, cost efficiency, operator-friendliness, and safety. Readers of this book will remain in awe and admiration for the recognized product developments, many of which came during his distinguished service as chief engineer.

I was grateful to Herb for bringing forth the Three Beauties—the Fleetline 66, 77, and 88—for opening the diesel engine era, and for his involvement in developing the Super 55, the 1800, the 1900, and all the others that followed.

I can truthfully state that Oliver Corp., as a prideful subsidiary of White Motor Corp. during my term, was the most profitable and shareholder-appreciated White Motor operation. This book tells how this was accomplished and how Oliver earned its coveted product leadership. I thank Herb on behalf of all the Oliver personnel, dealers, suppliers, and loyal customers everywhere. He is a "10" in my book.

Sam W. White Jr.
July 1996

PREFACE

It was a special privilege to have been a part of agriculture and agribusiness through most of the 20th century. Our main goals were to keep the cost of food as low as possible and to reach the areas of the world where people were hungry.

Growing up on a farm in Kansas, I was not happy with the quality and function of farm equipment in the 1920s and 1930s. It seemed necessary to make modifications to each piece of equipment. For example, on our Farmall tractor we added steps and other ways a driver could rest his feet to reduce body fatigue.

Other good examples of changes that were needed were the two McCormick-Deering reapers our family purchased from their last production run. The first reaper was already assembled with soft steel bolts and capscrews. The bolts and capscrews soon wore down to the point that looseness caused early failures. The frame was put together with nuts and lock washers, but the washers soon broke from the looseness. The bearings for the main shafts were made of cast iron. The combination of steel shaft running in a cast iron bearing was not the best.

We purchased the second unit unassembled and assembled it on our farm. In all of those places where we had experienced problems, we made modifications to correct the problems. We used heat-treated bolts and capscrews, and left out the lock washers, adding a second nut instead. The first nut was tightened to nearly the maximum strength of the heat-treated bolt. The second nut was tightened securely to the first nut which became a locknut. We removed the cast iron bearings, aligned the shafts and poured babbit material in the place of the cast iron. This second unit lasted for thousands of hours cutting flax in eastern Kansas.

The Great Depression of the early 1930s caused many changes in my life. My high school studies were all of the commercial type because I was planning to be a farmer. I graduated in May of 1934 and went into partnership with my father and youngest brother on the farm. The next two years were devastating ones for me on the farm. There were 10 days in July when the temperature never got below 100 degrees F day or night. There had been very little rain all year. Farmers were selling most of their livestock because of the shortage of water. My brother-in-law, George Weatherbie, was in the trucking business hauling grain to market and livestock to the Kansas City Stockyards. He had contracted to haul many loads of livestock that kept him busy from Sunday noon to Wednesday morning. George drove the truck when it was loaded and I slept in the truck's cab. Then I drove the empty truck while George slept.

Because there had been a potato crop failure in our area, we drove through the Kansas River area near Lawrence, Kansas, where potatoes were advertised at ten cents per bushel if we picked them up. We cleaned the truck and lined the bed with tar paper to take a load of potatoes to our home town to sell.

As I picked up potatoes, I could see the beautiful University of Kansas buildings of native stone and red tile roofs on a large hill. I asked George if we could stop by the engineering department. I changed to a pair of clean overalls and made the visit. Professor Jones, Acting Dean of Engineering, was very supportive of my desire to get an engineering degree with a goal to design better farm equipment. He directed me to Mrs. Palmer who was in charge of employment for students. She assured me that I could find employment through requests that came to her and youth employment programs. I graduated with a Bachelor of Science degree in mechanical engineering with majors in design and power plant.

When I graduated in 1941, the farm equipment companies were not hiring young engineers, so I accepted an offer from DuPont to design factory machinery in Wilmington, Delaware. But I was directed instead to Remington Arms Co., a division of DuPont, in Bridgeport, Connecticut, where I became an ammunition engineer. I was frozen to my job for the war effort until mid-1944. By this time, food and farm equipment had become the top national priority, so I decided to pursue my special interest in farm machinery engineering. I had many interviews and accepted a position as design engineer at

The late T. Herbert Morrell (right) proudly takes in one of the tractors he helped create during his tenure at Oliver. ***T. Herbert Morrell collection***

the Oliver Farm Equipment Company in Charles City, Iowa. In 1950 I was promoted to assistant chief engineer and then to chief engineer in 1951, a position I held until 1965 when I moved to the Chicago office as coordinator of outside products. I continued to be involved as a consultant in further developments and improvements of the Fleetline Tractors until 1970.

When I was first asked by members of the Hart-Parr Oliver Collectors Association to write this book, I declined. But they said that the development of the famous Fleetline 66, 77, and 88 tractors is an important story, and since Oliver is no longer a company, I should write a book about it because I know more of the details than anyone else alive today.

The famous Oliver Fleetline Tractors established a new standard for the industry with many innovations such as practical and economical diesel tractors, independent power take-off, easy riding seat, and hydraulics controlled by electrical circuits. This book tells the story of the development of the Fleetline Tractors, the 66, 77,

T. Herbert Morrell (center) seated in the Engineering Department, Oliver Charles City Plant, while selecting Timken tapered bearings in 1961 with Ernie Williamson (left) of the Oliver Purchasing Department and Bob Morgan of Timken Bearing Co. *C.J. Gibbs*

and 88, and their successors. It is a look behind the scenes at the people involved, the design criteria, the engineering considerations, the testing, and the problems, along with their solutions.

T. Herbert Morrell
June 1996

AUTHOR BIOGRAPHY

T. Herbert Morrell was born and raised on a farm near Blue Mound, Kansas. In addition to farming, his family built silos, water reservoirs, barns, and other farm buildings. In the mid-1920s they maintained natural gas wells in the area. He went to a country grade school, attended high school in Blue Mound, and earned a Bachelor of Science degree in mechanical engineering from the University of Kansas in 1941. His first position was as an ammunition engineer for Remington Arms Co. in Bridgeport, Connecticut, during World War II. In 1944 he began working for Oliver Corp. in Charles City, Iowa, as a design engineer. He was promoted to assistant chief

T. Herbert Morrell (center) was the first banquet speaker for the Hart-Parr Oliver Collectors Association. After his formal talk, he stayed for an hour and a half answering questions. The collectors asked him to write a book. ***Blanche Morrell***

engineer in 1950 and to chief engineer by 1951. His title was changed to senior chief engineer in 1963 when crawler tractors became part of his domain.

During his time with Oliver's engineering department, Morrell played key roles in the development of several innovative Oliver products and features, including the Fleetline tractors, power take-off, and the testing of the experimental XO-121 tractor.

From 1965 to 1970 he was coordinator of outside projects at Oliver's corporate headquarters in Chicago. He was vice president of engineering with Owatonna Manufacturing Co. until 1977 when he became a management consultant for design engineering and safety. Morrell has been an expert witness for approximately 150 lawsuits. A member of the Society of Automotive Engineers and a Fellow (Distinguished Member) of the American Society of Agricultural Engineers, he was especially interested in the safety of agricultural equipment, and served for many years on the safety standards committees of the SAE, ASAE, and the American National Standards Institute. Even though his health was declining, he worked very hard to finish this book. The manuscript was completed on September 29, 1996, and he died three days later.

CHAPTER ONE

Company History

The Oliver Corp. roots can be traced back to Hart-Parr, a company founded by Charles W. Hart and Charles Parr. Hart was born in Charles City, Iowa, in 1872. After one year at Iowa State University of Ames, Iowa, he transferred to the University of Wisconsin at Madison. Charles Parr was born March 18, 1868, on a farm near Dodgeville, Wisconsin. He met his future partner, Charles W. Hart, at the University of Wisconsin at Madison. In 1895, while still students at the University, Hart and Parr started experimental work on gasoline engines. Two years later they formed the Hart-Parr Co. on April 29, 1897, in Madison, but they had difficulty getting financial backing there. Hart's father and Charles City bankers C. D.

Charles W. Hart (1872-1937) was born in Charles City, Iowa. He and Charles Parr formed a partnership while still students at the University of Wisconsin. They will be remembered as the inventors of the first successful commercially produced agricultural tractor. Hart was the idea man of the two partners and left Hart-Parr in 1917 for a career in the oil refinery business in Montana.
Floyd County Historical Society

Charles H. Parr (1868-1941) was born near Dodgeville, Wisconsin. He and Charles Hart produced the first commercially successful tractor. Parr was the detail man, the engineer who developed the detailed plans for their inventions. Except for one year as chief engineer of the Elgin Street Sweeper Co. in Elgin, Illinois, Parr spent his entire career with Hart-Parr and its successor, Oliver Farm Equipment Co.
Floyd County Historical Society

and A. E. Ellis persuaded them to move the Hart-Parr Co. to Hart's hometown, Charles City, Iowa.

Both Hart and Parr were very civic-minded and active in Charles City affairs. For example, J. E. Waggoner was an Iowa State University graduate in mechanical engineering with a special interest in internal combustion engines. When he was hired to work in the Hart-Parr engineering department, no tractor design assignments were available. Instead, he was assigned the job of designing a suspension foot bridge over the Cedar River to connect the main residential district of Charles City with the athletic fields across the river. The foot bridge was placed on the National Register of Historic Places in 1989.

During World War I, Hart-Parr got an order to make artillery shells for the U.S. Army. But steel was in such short supply that they had to make it themselves. By the time they had assembled the tools, machined all the parts and were ready for production, it was 1917 and the government canceled the contract because they had already stockpiled enough artillery shells for the rest of the war.

Hart left the Hart-Parr Co. that year because of a conflict with the Charles City bankers C. D. and A. E. Ellis. He moved to Montana where he established the Hart Refinery and registered patent No. 1,458,936 on June 19, 1923, for a "method of harvesting grain." His invention was a header barge which cut and collected the heads of the grain crop. The Hart Refinery in Hedgesville, Montana, supplied gasoline and other petroleum products to approximately 10 filling stations that he owned. With a partner Hart set up a larger refinery in Missoula, Montana, and continued in the oil refinery business until he died March 24, 1937. He was buried in Riverside Cemetery in Charles City. His eulogy was read by his former partner, Charles Parr.

Parr remained with Hart-Parr and the new company, Oliver Farm Equipment Co. except for a short period in 1923–24 as chief engineer of the Elgin Street Sweeper Co., Elgin, Illinois. He died in June 1941, and was also buried in Riverside Cemetery.

The Hart-Parr Co. started producing stationary engines in a small brick building. Approximately 305 stationary engines were built from 1898 to 1904. The more powerful two- and four-cylinder engines were manufactured for stationary use during 1921 to 1929.

Once in the early 1950s I was taking Alva Phelps, president of Oliver Corp., along with some other Oliver executives, on a tour of the Charles City tractor plant. We saw a strange part being machined and asked what it was. The blueprint called it a Stationary Engine Casting, a cylinder for a stationary engine that had not been manufactured since 1904. They were being machined to fill a repair order for 50 parts to be shipped to Australia!

While stationary engines were powerful and durable, farmers needed their machinery to move through the fields. So Hart and Parr, while they were producing their stationary engines, were designing movable machines they called "gasoline traction engines." The first, Hart-Parr No. 1, was manufactured in 1902 and sold to an Iowa farmer who used it for about five years. Hart-Parr No. 2 was built in 1902 and sold in 1903. Hart-Parr No. 3, built in 1903, was used by the original owner for 17 years and is now in the Agricultural Collection of the National Museum of American History at the Smithsonian Institution Museum in Washington D.C.

Hart-Parr gasoline traction engine production really took off in 1903 with 14 Model 17-30s and 21 Model 22-45s. The first number indicates the rated draw bar horsepower (17 or 22) and the second is the rated belt horsepower (30 or 45). These models were continued in production until 1906. In 1906 the valve gearing was changed from pushrod type to rotary valve, and the tractors were re-rated 22-25 in 1908. The Model 30-60 was produced from 1911 to 1918 and nicknamed Old Reliable. A detailed listing of all tractors manufactured by Hart-Parr and its successors is given in Appendix A.

In 1964 an Old Reliable, built in 1913, was acquired by the Oliver Management Club. They gained permission to construct a steel and Plexiglas display shelter on the Floyd County Court House lawn where it stood until the Floyd County Historical Society was organized. The Floyd County Museum in Charles City now has three Hart-Parr tractors that were manufactured before 1920, the Old Reliable (30-60), a Hart-Parr (20-40), and a Little Red Devil (15-22). The latter two were purchased in the summer of 1996 after a Bring the Tractors Home fund drive. On May 18, 1996, the American Society of Mechanical Engineers declared the Hart-Parr

This 1913 Hart-Parr 30-60, nicknamed Old Reliable was purchased in 1964 by the Oliver Management Club and displayed in front of the Floyd County Court House in Charles City, Iowa. It is now in the collection of the Floyd County Museum.
T. Herbert Morrell

tractor a National Historic Mechanical Engineering Landmark. This was based upon research by John E. Janssen that determined Hart-Parr to be the first successful tractor plant in the USA.

It was Hart-Parr's sales manager, W. H. Williams, who thought "gasoline traction engine" was too long and reduced it to "tractor." Williams was not the first to use the term. Geo. H. Edwards of Chicago had used it in patent No. 425,000 issued in 1870, but his invention was a steam engine that was never produced commercially so the term never caught on. Williams' use of the word tractor did catch on and has since become an accepted word in the English language.

A railway was needed to connect the Hart-Parr plant with Illinois Central and Rock Island Railways for transporting the tractors, so the Charles City Western Railway was organized in 1910. It was capitalized at $300,000. The directors were C. W. Hart, C. H. Parr, A. E. Ellis, C. D. Ellis, E. M. Sherman, N. Frudden and F. W. Fisher. During the first directors' meeting, C. W. Hart was elected president of the board. The Charles City Western Railway was originally from Charles City to Rockford, Marble Rock, and Greene. Later, it was extended to Colwell, Iowa.

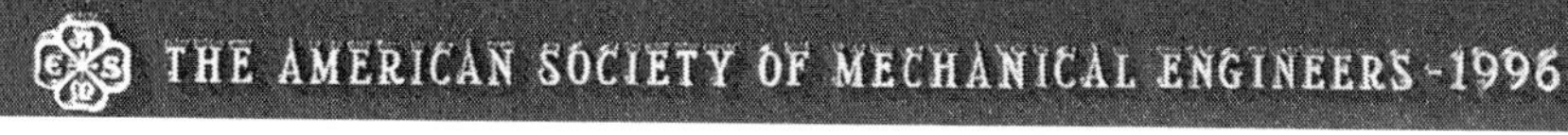

This National Historic Mechanical Engineering Landmark plaque declares the Hart-Parr No. 3 tractor as the first successful tractor in the world powered by an internal combustion engine. ***T. Herbert Morrell***

Thede Rowley was our next-door neighbor when we first moved to Charles City in 1944. He had retired after spending most of his life working for Hart-Parr Oliver and had many interesting stories to tell. I will always remember one story that he told. During his last few years at Oliver, Mr. Parr was referred to as the Old Man. The plant had a very high-capacity jack that was used for lifting big machines which also was referred to as the Old Man. Thede and others were involved in repairing some factory machinery and a new employee was asked to get the Old Man as it was needed to lift the machine. He came back with Mr. Parr instead of the big jack.

C. E. Frudden was chief engineer of Hart-Parr for many years until approximately 1920 when he left to work for Parret Tractor and later became chief engineer or director of engineering of Allis-Chalmers. It was my pleasure to visit with C. E. Frudden at a meeting of the Society of Automotive Engineers in Milwaukee, Wisconsin, after I became chief engineer of Oliver in October 1951. Our conversation centered upon current industry problems and standards.

A. H. Witt was another important figure in the early Hart-Parr organization. Shortly after he was employed by Hart-Parr, he entered the accounting department. In September 1917, he was made assistant secretary and assistant treasurer. He became Secretary in July, 1920 and continued as assistant treasurer. He held this position until he became comptroller of the Oliver Farm Equipment Co. following the merger.

Hart-Parr Innovations

Hart-Parr was the first to manufacture a practical internal combustion tractor engine. This was a better alternative to steam power on the farm because the large quantities of water and fuel required made steam expensive. Hart-Parr tractors required only low-cost liquid fuels. This innovation started a new tractor trend, and the Charles City Hart-Parr plant became the first successful tractor plant.

Not only were Hart-Parr tractors the first of their kind, the company continued in that tradition of innovation with many creative ideas through the years. Most Hart-Parr tractor engines could burn low-grade, low-cost fuels, such as kerosene or distillate. The engine

Hart-Parr manufactured stationary power units in Madison, Wisconsin, and continued in Charles City, Iowa, from 1897 to 1904. Mike Shanks owns this restored stationery engine and lives on a farm near Nora Springs, Iowa, which was originally purchased by his grandfather, Winfield Hart, brother of Charles Hart. It is believed that this farm was used by Hart-Parr for experimental purposes. *Floyd County Historical Society*

was started on gasoline and, after it was warm, could be switched to lower-grade fuel. In order to ignite the fuel when the engine was idling, the lower-grade fuel was heated by shunting it around the exhaust manifold. Under full load conditions, the low-grade fuel was not shunted because the engine heat was great enough to keep the fuel warm to assist the ignition. Some of these engines had water injection for anti-knock when burning kerosene. This reduced the cost of operation to half since gasoline was quite expensive.

Hart-Parr power ratings were quite conservative. For example, the 18-36 Model developed 32.25 horsepower on the drawbar and 42.85 on the belt. Such a rating provided surplus power of 79 percent and 19 percent, respectively. The tractor engine speed was governed to increase pulling power at lower than maximum engine speed, which allowed the tractor to carry the heavier drawbar loads through the tough spots.

This Hart-Parr No. 1 tractor was manufactured as a prototype, the first gasoline tractor. It was purchased by an Iowa farmer and used for several years. Neither this tractor nor the second Hart-Parr produced is still in existence.
Floyd County Historical Society

The company boasted about having enough power for large threshers and combines. Hart-Parr tractors were very successful in the large wheat fields of the USA, Canada, Australia, New Zealand, South America, Africa, and Asia. Smaller tractors, such as the 12-24 model, were popular in France, Italy, Hungary and other areas of Europe. So, Hart-Parr became the first American tractor company with substantial foreign business.

Hart-Parr developed the first:

- Oil-cooled engine, which had advantages because the oil would not freeze in cold weather and damage the engine.

- Valve-in-head engine which was more efficient and had longer life than the L-head engines with the valves in the crankcase. The valve-in-head engine was easier to service and provided better fuel economy.

- Force-fed lubrication, which provided more adequate and continuous engine lubrication.

- Multi-speed transmission which added to the versatility of tractors so they could do more agricultural functions that had been done by horses. The transmission and final drive gears were over-designed, resulting in good reliability and long life.

Hart-Parr offered quite a selection of wheels for angle, spade, and spike lugs. In fact, the 12-24, 18-36, and 28-50 models were offered with solid rubber treads on the front and rear wheels. Rubber treads were provided for maintaining roads and where conventional steel wheels and lugs were not permitted or were not practical.

The third of the original Hart-Parr tractors saw 17 years of use on the farm before being donated to the Smithsonian Institution Museum of History and Technology in Washington, D.C. in 1949. *Floyd County Historical Society*

The Hart-Parr 20-40 featured two forward speeds and was built from 1911 to 1914. This one was sold at auction in the summer of 1996. Thanks to a successful Bring the Tractors Home fund-raising campaign, the Floyd County Historical Society was able to buy it. *Jeff Hackett*

The Hart-Parr 30-60, the Old Reliable, in its heyday. Hart-Parr offered a wide selection of wheels for angle, spade, and spike lugs. In fact, the 12-24, 18-36, and 28-50 models were offered with solid rubber treads on the front and rear wheels. Rubber treads were offered for use where conventional steel wheels and lugs were not permitted or were not practical. *Floyd County Historical Society*

Oliver maintained a tie to its Hart-Parr roots with tractors by labeling their tractors Oliver Hart-Parr. This 18-27 Row Crop has Hart-Parr in large type with Oliver in small type above. It is outfitted with an Oliver planter.
Floyd County Historical Society

Hart-Parr introduced the first tricycle Row Crop tractor and copyrighted the name Row Crop. The higher rear axle clearance of the Row Crop tractors provided more versatility.

The independent power take-off (PTO) was first introduced on the Hart-Parr 18-36 in 1929. But it was way ahead of its time and there was not much interest in this early innovation. The implements, with the exception of the combines, did not require much power so that the advantage of the independent PTO was not apparent. The independent PTO became a very important part of the development of the famous Oliver Fleetline tractors introduced in the late 1940s.

In 1928, prior to rubber tires on tractors, Hart-Parr developed the Tip Toe rear wheels. The theory behind this development was for the rear wheels to penetrate the loose soil on top of the ground into more firm soil for traction. This gave the rear wheels adequate traction without compacting the soil on top. The Tip Toe rear wheels were very popular on the Oliver Hart-Parr tractors and the Row Crop 70 until rubber tires were introduced in the mid 1930s.

Hart-Parr and Oliver Contributions to Society

- Replacement of steam engine power for agriculture
- Successful internal combustion tractor
- First tractor production plant
- Kerosene-burning engine
- Oil-cooled engine
- Valve-in-head engine
- Tricycle Row Crop tractor
- Multi-speed tractor transmission
- Independent power take-off
- Forced-feed lubrication in tractor engines
- Gave the word “tractor” to the industry
- First foreign tractor business

Oliver Contributions

- Economical, practical diesel tractor
- Practical independent power take-off
- Mass production of six-cylinder tractors
- Tip Toe wheels
- Electrical control of hydraulics
- Equalizer brake pedals
- Double-disc brakes in tractors
- Low-pressure engine lubricating system
- Aluminized steel mufflers
- Tilt and telescoping steering wheel
- Four-wheel drive with Terra-Tires
- Wheel guard fuel tanks
- Certified horsepower
- Two-point hitch and lower link draft control
- Cast iron grill for Row Crop front stability
- Ridemaster seat
- Electric lights
- Bale Thrower
- Throw-away Raydex Plowshares

On this Row Crop tractor, an Oliver and Hart-Parr emblem are visible directly above the grill. Oliver's first Row Crop tractors were designed in 1928. Row Crop models used a tricycle or narrow-front design. The maneuverability, ground clearance, and ability to straddle rows made Row Crop tractors ideally suited to cultivation. *Jeff Hackett*

The Merger

James D. Oliver was the founder of the Oliver Chilled Plow Works. He was born in Scotland on August 28, 1823, and came to the USA when he was 11 years old. He worked at many different jobs, including a blacksmith's helper. In 1855 he bought interest in his first foundry where he manufactured Sulky Plows. The Sulky Plow was a riding plow with one- or two-bottoms that was normally pulled by three or four horses. At that time, cast iron was the material of choice for plows because it was inexpensive. But cast iron did not accept a polish and dirt would not move easily over the plow blade. Sometimes the plow could go only a few yards in gumbo soil and the operator would have to stop and clean the surface of the plow blade.

To solve this problem, Oliver developed a process of chilling a cast iron piece in the mold which hardened a chosen portion of the casting. This process created a very hard surface which could be polished and would allow dirt to pass over it easily.

James D. Oliver died in 1908. His son, Joseph D. Oliver, who had worked in the company as financial manager, became the chief executive officer after his father's death.

On April 1, 1929, the Hart-Parr Tractor Co. of Charles City, Iowa, merged with Oliver Chilled Plow Works, South Bend, Indiana, and Nichols and Shepard Threshing Machine Co. of Battle Creek, Michigan. The following month American Seeding Machine Co., Springfield, Ohio, joined the fold. Together these companies formed the Oliver Farm Equipment Co. The oldest of these, the Nichols and Shepard Threshing Machine Co. was founded in 1848.

Joseph D. Oliver was elected chairman of the board of the new company. Melvin W. Ellis, president of Hart-Parr, became president of the new company and continued in that position until January 1, 1931, when C. R. Messinger became president and Ellis became a vice chairman of the board. The corporate headquarters were located at 400 W. Madison in Chicago.

As president of Chain Belt Co. in Milwaukee, Wisconsin, C. R. Messinger had been intimately connected with the agricultural implement industry for 14 years, and his abilities as a manager gained recognition. He served on many boards of directors, including Milwaukee Gas Light Co., First Wisconsin National Bank, First Wisconsin Trust Co., First Wisconsin Co., Sivyer Steel Casting Co. of Chicago and Milwaukee, Interstate Drop Forge Co., Federal Malleable Co., and Stearns Conveyor Co. of Cleveland.

R. C. Rolfing was elected vice president and general manager of Oliver Farm Equipment soon after the merger in 1929. Under his leadership, the manufacturing plants were modernized to provide a generous supply of quality products for Oliver dealers. There have been many good comments about his contributions to the company.

W. A. Weed was elected vice president of Oliver Farm Equipment in late 1929. A. H. Witt continued as comptroller, a position that he held for many years with Hart-Parr before the merger. H. S. Lord, general sales manager, was well regarded for his enthusiasm and knowledge of the people in the Oliver branches and Oliver's products.

The Oliver flag on this 1930 Row Crop advertisement shows the four components of the 1929 merger that created the Oliver Farm Equipment Co.: Hart-Parr, Nichols and Shepard, Oliver, and American Seeding Companies. ***Floyd County Historical Society***

Dave E. Darrah was advertising manager. His motto for 1930 was "Tell 'em and Sell 'em—these two jobs we can do this year!" In 1931 Darrah became ill and left Oliver. Bert C. King took over as ad manager and designed the Oliver shield and flag. Through the efforts of these two men, modern literature on all of the products was available to all the dealers. A direct mailing was made to all farmers who owned Oliver products. The goal was to make them all-Oliver farmers. This all-Oliver prospects list was divided into eight general classifications—Canadian Wheat, Northwest Wheat, Southwest Wheat, Western Diversified, Eastern Diversified, Potatoes and Truck, Orchard and Southwest, and Cotton and South.

John L. Carpenter was treasurer. He had a good understanding of credits and collectibles. His training of branch employees and dealers about the financial aspects of the farm equipment business was excellent and helped the company succeed financially.

This merger was quite successful because the new company provided a full line of agricultural equipment, and Oliver Farm Equipment Co. became a leader in the agricultural equipment industry.

The spirit of inventive enterprise that marked the early activities of the Hart-Parr Tractor Co. continued with Oliver to produce many new developments to make the farmer's work easier, more efficient, and more profitable. Oliver was a leader in the development of products that would benefit society through cost effectiveness, interchangeability, safety, and other benefits to the owners.

Oliver Tractors for Russia

During the early 1930s, Russia wanted to purchase tractors and combines from the USA. The Russian representatives arrived a few days earlier than expected and Oliver was not ready to show the tractor designed to meet their specifications. To buy some time, Oliver contacted the Milwaukee and St. Paul railroad while the Russians were en route and persuaded the railroad not to stop the train at Charles City, but to continue to Mason City, which is about 31 miles west. Some Oliver representatives from Charles City met the train in Mason City, took the Russians to an elaborate breakfast,

This 1941 advertisement touts the Oliver 60 row Crop, which produced 15.17 drawbar and 18.35 belt horsepower when tested at Nebraska in 1941. ***Floyd County Historical Society***

toured some farms in the Mason City area, and then after lunch traveled to Charles City. Meanwhile, others worked all night so the tractor was ready to show to the Russians.

The Russians ordered 5,000 tractors, Oliver's model 28-44, and 2,800 Model F combines. Oliver's shipment of the tractors and combines started in 1930 and was completed in 1931. The sale of the tractors from Charles City and combines from Oliver's Battle Creek plant helped the company financially early in the Great Depression. After the Russian tractor order was filled, the Oliver tractor plant was closed except for a few people working in engineering on the Model 70 tractor design, Archie Howland in charge of accounting, and Jim Smith in charge of shipping and repair orders. Edna Ladd, who was secretary to the plant manager, was moved to the switchboard to take incoming calls which were primarily for repairs. One year she worked all summer without pay. She continued to work for Oliver for a total of 45 or more years.

Further Corporate Changes

In 1944 Oliver bought the Cleveland Tractor Co., Cleveland, Ohio, and the name changed from Oliver Farm Equipment Co.

This Oliver Industrial tractor equipped with dual rear wheels is preparing a roadbed. It is most likely based on the Model 50, a four-cylinder tractor built from 1937 to 1948 that was available only in the standard tread configurations. ***J.C. Allen and Son***

to The Oliver Corp. The acquisition of Cleveland, with its line of crawler tractors, expanded Oliver's focus from just farm equipment to include industrial products as well.

From 1944 to 1960, Oliver acquired several more companies, including Be-Ge of Gilroy, California, a hydraulic equipment manufacturer; Farquar & Iron Age of York, Pennsylvania, that made equipment for potatoes and conveyors; and Chris Craft Outboard Motors.

I recall a disheartening situation when I was privileged to review some confidential finance information. During the first quarter of 1958, the tractor plant at Charles City had accounted for several million dollars net profit. The Battle Creek harvesting plant was in its off season but still accounted for about $160,000 net profit. The Cleveland crawler tractor plant was breaking even. The other seven plants showed losses amounting to about three-fourths of the profit from the Charles City tractor and the Battle Creek harvesting plant.

In 1960 White Motor Corp. bought some of Oliver's plants, including the plant that manufactured tillage equipment in South Bend, Indiana; the hay equipment plant in Shelbyville, Illinois; the tractor plant in Charles City, Iowa; the harvesting equipment plant in Battle Creek, Michigan; and the seeding equipment plant in Springfield, Ohio. The name changed from The Oliver Corp. to Oliver Corp., a subsidiary of White Motor Corp. The remaining plants became a part of Amerada Hess. Lastly, in the early 1960s White bought the Cleveland line of crawler tractors and moved it to the Charles City plant. This, too, became a part of Oliver Corp.

In 1961, White Motor Co. purchased Cockshutt of Canada and made it a part of Oliver Corp. The next year, White Motor Co. bought Minneapolis-Moline, but operated it as a subsidiary of White Motor Co. separate from Oliver Corp. because of the US antitrust laws.

From approximately 1960 to 1966, Oliver was a major source of profit for White Motor Co. The farm equipment business went into a minor recession in the late 1960s and White Motor executives criticized Oliver, Minneapolis-Moline, and Cockshutt for not contributing enough to the overall company profit. But the farm equipment profit as a percentage of sales was greater than that of their truck division.

In late 1969 Oliver, Cockshutt and Minneapolis-Moline were integrated into one organization called the White Farm Equipment Co., Division of White Motor Corp. The division's headquarters were established in Hopkins, Minnesota. Over the next five years the individual company names disappeared and all of the colors were changed to the two-tone gray color of the White Models. After 1975 the names Oliver, Minneapolis-Moline, and Cockshutt no longer appeared on the tractors which were manufactured at Charles City, Iowa.

White Motor Co. experienced economic problems during the late 1970s. In December 1980, Texas Investment Corp. purchased White Farm Equipment Co., Division of White Motor Co., and operated it as a wholly owned subsidiary. In November 1985, selected assets of White Farm Equipment Co. were purchased by Allied Products Corp. In 1986, White Farm Equipment purchased selected assets from White Farm Manufacturing Ltd. in Canada. In May 1987, White Farm Equipment merged with another Allied subsidiary to form White-New Idea Farm Equipment Co. with headquarters in Coldwater, Ohio.

The next change in ownership was in June 1991 when AGCO purchased the Charles City plant. The next year, after 86 years of tractor manufacturing in Charles City, Iowa, the plant was closed. On October 29, 1993, there was an auction to sell what was left of the Charles City tractor plant. The plant was demolished during the fall of 1995. The power plant smoke stack was imploded December 8, 1995, and the rest of the power plant was torn down by January 1, 1996.

Although the company no longer exists, the Oliver brand is kept alive by a large number of dedicated Oliver customers and the Hart-Parr Oliver Collectors Association. The association had over eight thousand members in 2010 and membership is growing. Some of the collectors have as many as 100 historic and restored tractors.

The People Who Developed the Fleetline Tractors at Oliver's Chicago Office

Many people contributed to the development of the famous Fleetline Tractors during the 1940s and early 1950s, including those in Oliver's executive office in Chicago. Alva Phelps became chairman of the board of directors soon after joining Oliver in 1944. He was formerly employed by Saginaw Steering Gear Division of General Motors. The very efficient recirculating-ball steering gear was his invention. The principle of his invention was to use the lower resistance of steel balls rolling in grooves rather than conventional gears to steer a vehicle. The friction between the gear teeth and a worm gear in the conventional steering gear was far less efficient and less responsive than the recirculating ball. Because of his engineering background, Phelps was a strong booster of the Fleetline Tractor development program.

A. King McCord was president during the period of the late 1940s and until late 1954 or early 1955 when he resigned to become

Oscar Eggen, vice president of Oliver engineering, is driving a Model 66 Industrial tractor that is pulling a trailer load of sightseers during the 1952 Tired Business Men's Golf Tournament at Charles City. ***Carl Rabe***

Oliver Chief Engineer Herb Morrell attending a 1954 Society of Automotive Engineers meeting at Chicago's Edgewater Beach Hotel with business associate and friend Kip Recor, sales representative for Torrington Co. that supplied straight roller bearings to Oliver.
T. Herbert Morrell collection

president of Westinghouse Airbrake Co. Of all the chief executives, I enjoyed him most in meetings about products, engineering, future planning, and many other aspects of our company. He was down to earth, a good listener, and always appeared to be unselfish. His efforts were for the good of the company.

Oscar Eggen was Oliver's vice president of engineering starting in 1942. He had become chief engineer of the tractor plant in Charles City when the Oliver Farm Equipment Co. was formed. He had ample experience and good knowledge of farm equipment, particularly about tractors. He was full of good suggestions and had very good insight into future planning and defining product requirements. He resigned in 1954 and became a manufacturer's representative in the Los Angeles area.

Roy Melvin was vice president of manufacturing. He had been plant manager at Charles City at one time. His expertise was in industrial engineering. Roy had some health problems and resigned from his position in the executive office. After he recuperated, he came back to the Charles City plant as supervisor of the Methods Department, also referred to as the Industrial Engineering Department.

Herb Morrell, George Curtis and his replacement at Timken Bearing Co., and George Bird, Oliver's Charles City plant manager, evaluate a Timken bearing design in the early 1950s. In the background is Ida Smalley, the engineering department's secretary. ***Carl Rabe***

Joe and Merle Tucker were brothers in Oliver's executive office. Merle was vice president of sales. Joe resigned shortly after I started with Oliver. He had the idea to design a self-propelled combine, but his idea was neither accepted at the combine plant in Battle Creek, Michigan, nor in Oliver's executive officers. So Joe took his idea to Massey-Harris Co. in Racine, Wisconsin. The Massey-Harris combine was so successful that it became the standard for a long-lived harvesting system. Massey-Harris merged with Ferguson of England in the late 1950s and became Massey-Ferguson Co. Joe's next move was to the New Holland Co. of New Holland, Pennsylvania.

Black and White were two young fellows who traveled together to the sales branches on a public relations mission. Black left Oliver after a very short employment, but White was Samuel Walter White Jr. Sam's father Samuel W. White Sr., was on the board of directors. He was the Chicago investment banker who put the new Oliver Farm Equipment Co. together from the four family-owned businesses in 1929.

Sam White Jr. started as a trainee in 1939 at South Bend, Indiana. He spent 1942–46 in the United States Navy. After the war, he held several positions in the corporation, including president of Oliver International S. A., the company's export subsidiary. In 1960 he became president of the new Oliver Corp., a subsidiary of White (no relation) Motor Corp. During the last part of his Oliver association, he was White Motor's executive vice president of Farm and Industrial Equipment. He was very knowledgeable of the agricultural and industrial industries, and one of the best at seeking Oliver recognition in our industry.

J. Oliver Cunningham, a grandson of the founder Joe Oliver, was in Oliver's advertising department during the latter stages of the Fleetline tractor development. Later, he became the manager of the Memphis, Tennessee, sales branch. J. Oliver and I had fun winning at bridge during the evening when he came to Charles City to develop advertising information. He resigned from the company and moved to Arizona where he sold appliances and developed some good hobbies. We have exchanged Christmas greetings ever since and we look forward to receiving his unique and clever greetings each year.

These are only some of the many Chicago Office Oliver executives involved directly with the development of the famous Fleetline 66, 77, and 88 tractors.

Sam White Jr. (right), president of Oliver Corporation from 1960 to 1970, is shown with George Bird (left), plant manager of Oliver's Charles City plant from 1944 to 1961.
T. Herbert Morrell collection

Charles City Plant Operating Committee

Oliver's Charles City plant manager George W. Bird operated an Old Reliable in Montana when he was a youth in the mid 1910s. He looked at the tractor name plate and noted that it was built by Hart-Parr in Charles City, Iowa. He decided that he would rather build the tractors instead of driving them. He came to Charles City and began a very interesting life with Hart-Parr and Oliver.

The Charles City plant had an operating committee consisting of the chief supervisors and some assistants that met each Wednesday at noon. The food was catered from the St. Charles Hotel or a local restaurant and served by committee members on a rotating alphabetical basis. Reports of various phases of plant operation were presented and discussed.

In addition to those pictured, several others were on the operating committee during the period of the design and manufacture of Oliver tractors. Gordon Atherton became foundry superintendent when Brunsman retired. Ernie Williamson became chief purchasing agent after Kuehn's retirement. Frank Praytel became plant manager in 1961 after Bird's retirement. Dick Bennett took over as chief metallurgist when Carbaugh retired. After Ed Kroft went to the White Farm Equipment office in Hopkins, Minnesota, John Culbertson became personnel manager and retained that position until the plant closed in late 1992.

This operating committee was one of the hardest working and most dedicated groups in the industry. The cooperation among the various

George W. Bird is sitting on the front wheel of an Old Reliable tractor he operated as a youth in Valier, Montana. He came to Charles City and was hired by Hart-Parr. He had many positions, working up to plant manager in 1944, a position he held until 1961 when he retired. *Floyd County Historical Society*

The Charles City Plant Operating Committee was responsible for the development of the famous Oliver Fleetline 66, 77, and 88 tractors. This photo was taken December 21, 1951. (outside row, from left) E. A. Brunsman, foundry superintendent; Ralph Battey, production control; Herb Stokes, service manager; Bill Kuehn, chief purchasing agent; Archie Howland, supervisor of accounting; Lyle Lenth, maintenance superintendent; Roy Melvin, methods superintendent; Herb Morrell, chief engineer; George Bird, plant manager; Melvin Finch, general factory superintendent; Lloyd Maby, production control supervisor; Merle Tucker, vice president of sales in Oliver's Chicago office; Jim Martin, methods department; George Taylor, accounting department; John Rockufeler, supervisor of part machining; and Bob Burgraff, suggestion secretary. (inside row, from left) Russell Elliott, production control and special assignment; Merle Hicks, sales order manager; Dale Tower, maintenance department; Phil Carbaugh, (nearly hidden) chief metallurgist; Lloyd Boyle, shop supervisor; Ed Kroft, personnel manager; Tony Obermeier, supervisor of assembly line and later shop superintendent; Leland Hartwell, chief inspector; Pete Burns, assistant chief engineer. *Carl Rabe*

departments was outstanding during the time the Oliver Fleetline Tractors were designed, developed, and put into production. We used some efficiency and management techniques, such as keeping the various departments informed of new product developments, which have been reported in recent years as if they were new.

Oliver Engineering Personnel

The planning of the Fleetline tractors was initially the responsibility of Vice President of Engineering Oscar Eggen, Chief Engineer Louis Gilmer, Assistant Chief Engineer Milford Stewart and Supervisor of Experimental Engineering Tom Martin. During the mid 1940s Bob Butler was hired as service manager and John Dorwin as assistant service manager. The service department, which was under the engineering department, was responsible for the technical publications, such as operator's manual, parts catalogue, service manual, and service bulletins.

John Selim was hired in Engineering to be the industrial tractor sales representative in 1945. He also worked with outside suppliers to provide loaders and other attachments for the industrial tractors. John was transferred to Cleveland, Ohio, to work with the industrial sales department during the late 1940s.

Over the next 20 years there were a few changes in the Charles City plant engineering leadership. In 1946 Milford Stewart went to South Bend, Indiana as chief engineer of the Oliver plant. Tom Martin became the assistant chief engineer of the Charles City plant, Pete Burns took his place as supervisor of experimental engineering, and Homer Dommel became his assistant. In 1948 Tom Martin went to Ohio as chief engineer of the Springfield plant. Bob Butler took his place as assistant chief engineer; John Dorwin became service manager replacing Bob Butler, and Bill Brown became his assistant.

In 1950 Bob Butler went to South Bend to manage a new program on stationary power units. I replaced him as assistant chief engineer. When Louis Gilmer died in 1951, I replaced him as chief engineer, Pete Burns became assistant chief engineer, and Homer Dommel became supervisor of experimental engineering. In 1954 Pete Burns left the company to be a member of the product planning department of Ford's tractor and implement division and Ron Ronayne became assistant chief engineer.

In 1965, I moved to the Chicago office to become coordinator of outside products, those products that are manufactured by another company for Oliver to sell through its dealer organization. Ron Ronayne became chief engineer of the Charles City Plant and Bob Prunty became his assistant.

Some of the principal design engineers during the development of the Fleetline were Wendell Neland on sheet metal, Walt Roeming on engines, Charles Van Overbeke on engines, Don Kinch on transmissions, Chuck Ruhl on the new Hydra-Lectric hydraulic system, Joe Pieper on front frame and front end in general, and I was assigned the rear end, power take-off, pulley, and general accessories behind the front frame.

Roy Sobolik was in charge of the drafting department which made and checked all of the drawings. Among the draftsmen was Harrison Lambkin who did the design and drawing checking. Harrison was an unusual, but accomplished person. He had polio at age 4 and was handicapped. He educated himself through correspondence courses and excelled in mathematics. He was George Bird's brother-in-law.

The specifications department was responsible for filing tractor drawings, providing part numbers, compiling and typing the bills of material from which the tractors would be manufactured and assembled. Ray Hennagir was the supervisor of this department at the outset of the Fleetline program. He retired shortly after the experimental models were manufactured. Morrie Thelen, his assistant, became the supervisor and continued until the plant closed.

As assistant methods superintendent under Roy Melvin, Art Munson worked on special tooling and dies for forming sheet metal and machines. Bob Lockhart, also in the methods department, worked on jigs and fixtures to machine castings and other parts.

The experimental engineering department did all the testing of the tractors and assemblies, such as the transmission, pulley mechanism, power take-off, and hydraulics. Keith Minard, Gene Lockie, and Max Denham were excellent electrical dynamometer operators and also good at setting up field tests. Each tractor model had to be tested at the University of Nebraska in order to sell that model in the state of

Nebraska. Max Denham accompanied most of the tractor models to Nebraska and assisted with the tractor tests. Charlie Adams was an unusual person and was gifted in visualizing a test fixture that would test some part or assembly of parts. He had experience at the University of Nebraska and was quite familiar with various types of test equipment before coming to Oliver. Joe Roland had a good personality for managing a remote or local tractor field test program.

One of the most valuable employees in engineering was Ida Smalley, secretary to the chief engineer of the Charles City plant. She had a firm knowledge of what was happening in the department and kept me informed of her evaluations. On a big day for mail, she would scan what we received and categorize it into as many as five piles—junk mail, reports with some notations by her, information only, to be answered, and emergency action required. She was a very fast typist and did all of the letters going out of engineering.

When she retired in 1961, she had been secretary to the chief engineer of the Charles City tractor plant for at least 42 years. I knew that it would be difficult to replace her. I sent two young ladies to Ida to be trained. However, this was unsatisfactory because she made such great demands that both of them refused to accept the position.

Mildred Block, who worked in the accounting department, was a good prospect. I decided to let Ida retire and then Mildred and I would work together for her training. Mildred soon became a very efficient secretary.

One could not ask for a more dedicated group of people to do major design of tractors and attachments.

Suppliers

Oliver made every part that was economically practical and feasible, and purchased the parts or assemblies that were not. Suppliers of standard and special parts made important contributions to the development of the Fleetline 66, 77, and 88 tractors. Manufacturers of large production quantities of standard parts are specialists in their own businesses. A good example is the tapered roller, roller, and ball bearing industry. It would not be economical for a farm equipment manufacturer to make bearings, hardware, and many other high production standard type parts for their own low production requirements.

(Bill) Kuehn, Oliver's chief purchasing agent, had a big job in coordinating all of the outside purchased items. These included materials such as steel, coke for the foundry, pig iron, scrap iron, and thousands of standard items. There was a radio program in the 1940s and 1950s called "Mr. Keene, Tracer of Lost Persons." Sometimes Bill was jokingly called "Mr. Kuehn, Tracer of Lost Purchases." Bill was never happy about such a remark because he and his department did their best to manage inventory and have purchased items delivered on schedule. One of their goals was to have only a three-day supply of engines and no more than a five-day supply of tires on hand.

Oliver's purchasing and engineering departments entertained our suppliers at the Tired Business Men's Golf Tournament, a traditional Charles City Country Club weekend. Other Charles City businesses also invited guests. It was held the second weekend of June nearly every year from the late 1940s to the early 1960s. The guests arrived Thursday or early Friday morning for a business call. Friday afternoon was a practice round of golf for the guests. Friday evening was a social outing. On Saturday there was an 18-hole handicap tournament. A floor show was arranged for Saturday evening.

Sunday morning was a time for "crazy golf." A golfer may have been required to hit the ball from a swinging barrel or through the opening of a large tractor tire located on the fairway or to putt through a fence about 2 feet in diameter around the cup with openings that were not much larger than a golf ball. Most guests went home late Sunday, but some stayed for business calls on Monday. We felt that the weekend was truly a Tired Business Men's Golf Tournament because, if a guest wasn't tired when he arrived, he would be tired when he left Charles City.

Sales and Service

Sales and service functions were extremely important. The entire country was divided into sales branches, each of which was divided into territories. Each year new territory managers were sent to Charles City to learn all of the sales points of Oliver tractors. Each fall there was a school for branch sales managers who in turn trained the territory managers. Branch and territory managers were Oliver employees, but some independent distributors were also included in the sales and service schools. Coop Federee from Montreal, Canada was a good example. Special sales and service schools were held for this Quebec distributor.

Service schools were generally held during the third week in January. The branch service managers were sent to Charles City for a week's training on anything that was new in the tractors and a refresher course on the older units. These branch service managers were then to hold service schools for the dealer's mechanics in each territory manager's area.

If there was something new during the year, plant representatives were sent to the branches to conduct special sales and service schools. The Charles City plant had representatives assigned to each branch through the Branch Contact Program. The purpose was to provide a personal contact in the plant for each branch. The plant representatives traveled to the branch area to review anything new from the plant and learn of the problems so they could report back to the plant on possible improvements. Each team consisted of about five representatives, including a captain from engineering and an alternate member.

My team covered Harrisburg, Richmond, Memphis, and Columbus, Ohio, branches. Harrisburg was my branch. When the Hydra-Lectric unit was introduced, I spent one week in Harrisburg conducting a sales and service school for Harrisburg personnel in 1949. In 1950, I spent a week on sales and service for Seabrook's personnel in New Jersey during early March. This school was the result of Seabrook Farms purchasing about 69 Oliver diesel tractors. Each contact representative spent one week in the spring during tillage and planting season, one week during the cultivating season, and one week during the harvesting season. The contact teams were effective from about 1948 to 1952 when the program ended. After that engineering representatives were sent only as needed.

Most sales of Oliver tractors were made by its agricultural and industrial dealer network. Special tractor sales were handled by the tractor plant sales representatives. Oliver tractors were shown at most state, county, and other local functions such as in parades.

Oliver Facilities

The Charles City tractor plant occupied 65 acres of which 17 acres were under roof. For its time, the plant was one of the most up-to-date tractor plants in the world and underwent many renovations over the years to keep it modern. A good example was the 100,000-square-foot addition in 1948 to house a new assembly line, paint system, casting machining, and conveyorized handling of castings from the foundry. The foundry was remodeled to increase its capacity. This new addition was essential for the production of the new Oliver Fleetline 66, 77, and 88 tractors.

Foundry

The foundry in Charles City was updated several times during its life. The additions in 1948 provided for the extra production of the Fleetline tractors and for conveyor belts within the foundry for delivering castings to the machining area. This foundry was very flexible in providing the necessary variety of cast iron. One modernization of the foundry added hoods over the casting cooling area of the foundry conveyors to collect the smoke and fumes from the cooling castings, making the foundry one of the cleanest in the world. In fact, the foundry was still one of the most modern when it was closed in 1992.

An electric-powered furnace was installed during the mid-1950s to provide special castings, such as exhaust manifolds, which needed

This billboard greeted visitors to Charles City, Iowa, in 1956. The photo was taken by the author's brother-in-law. ***Harold W. Snyder***

The Oliver tractor plant at Charles City is shown in this aerial view. Engineering's oval test track is in the top center of the photo. ***Carl Rabe***

certain alloy additions to control growth under hot temperatures of the exhaust. The alloy chips from machining steel were used for some of the required alloys. These chips would have burned up in the regular cupola, but they could be loaded into the electric furnace when the furnace was cold and would not burn when the electric power was turned on.

Casting Machining

When the castings cooled, they were removed from the mold and placed on a conveyor. The conveyor transported the castings to a cleaning area where fins, excess sand, and so on, were removed. When the castings were clean, they were placed on the conveyor which transported them through a primer paint booth and then into the next building to the casting machining area. Castings were removed from the conveyor adjacent to the machinery to do the machining. After machining, the castings were placed back on the conveyor and taken to the area to be assembled into a subassembly

This photograph of the foundry at the Charles City plant was taken in 1956.
Harold W. Snyder

or moved onto the main assembly line. Some of the smaller castings were placed into a skid container and then parked by the assembly line near where they would be used.

Steel Machining and Stamping

After the 1948 expansion, steel machining for gears, shafts, axles, and all other steel parts was done in the area where all of the machining and the assembly line existed prior to the 1948 expansion. Chips from the alloy machining of gears were saved and used in the foundry's electric furnace.

Oliver developed a highly sophisticated gear manufacturing department. Gears were crown-shaved so that the center of the gear teeth was slightly thicker than the ends of the teeth. Crown-shaving provided for quick wear-in of the gears, resulting in accurate tooth contact with mating gears.

One building was devoted entirely to forming sheet metal for instrument panels, grills, wheel guards, hoods, and side panels.

The tractor assembly line before the 275-degree bake oven. Some parts, such as tires and batteries, could not withstand this high temperature, so they were installed after the bake oven. ***Carl Rabe***

New Assembly Line

The new assembly line was conveyorized, which provided smooth movement of the tractors being assembled. In fact, the assemblers could ride on the conveyor along with each tractor to designated points to finish their operations. They could then walk back to the beginning of their stations and start assembling the next tractor. The speed of the conveyor was timed for the assembly of so many tractors per day. Subassemblies, such as power take-off, pulley, and hydraulics were delivered to the main assembly line at the points where they were added to the tractor assembly.

Computer Control of Production and Assembly

We started assembling tractors using a computer system with the Supers in 1954. Key-punched cards provided information for parts and assembly. For example, the type of transmission would be indicated by a number signifying the parts that were to be assembled in that tractor. This provided information for the foundry, machine shop, and assembly line to complete the tractor. Another key-punched card showing the recommended list price accompanied each tractor to the dealer.

Another view of the assembly line taken in 1956. *Harold W. Snyder*

The tires, wheels, and wheel weights were assembled, and then the assemblies were attached to the tractor. These assemblies required cranes to position them on the tractor. *Carl Rabe*

The final assembly line was the place where the tires, wheels, batteries, and other low-temperature items were installed. The tractors were then started, the transmission operated, and a general inspection performed. Next they were driven on the rear-wheel power test rollers to make certain that they had the prescribed minimum drawbar power. ***Carl Rabe***

Painting

Paint companies related that the quality of paint on Oliver tractors was equal to or higher than the paint on U.S.-manufactured automobiles. The main assembly line passed through the paint booths. The paint was sprayed on the tractor from both sides. Both sides had a waterfall curtain which collected excess paint mist. The tractors then entered the oven to bake the paint. The oven temperature could be regulated up to 275 degrees F. It was important to adjust the temperature, depending upon the size of tractors and how fast the assembly line traveled. Perishable items, such as batteries and tires, had to be added to the tractors after the oven bake operation.

Tractor Testing

The testing of tractors by Hart-Parr is not well-documented. Apparently, Hart-Parr engineers and dealers observed and researched the needs for tractors and then Hart-Parr manufactured tractors with extra life to meet the customer requirements. While it is not clear what testing was used by Hart-Parr, the testing of the Oliver 60, 70, and 80 models was quite elaborate.

The testing of the 66, 77, and 88 followed some of these established test procedures with many new test exposures added. In the fall of 1944, six prototype tractors of each model were assembled, along with extra component assemblies such as engines, transmissions, differentials and final drives. Four of these were sent to Arizona where the tractors could be tested in fields for as long as 24 hours per day. Two prototypes of each model were retained in Charles City for tests by local farmers, for laboratory tests, and for redesigning parts as a result of the tests.

Deep plowing in the irrigated Arizona fields required near maximum horsepower, and the 12 prototype tractors were tested there through most of 1945 and then returned to Charles City for disassembly and evaluation of wear, strength, and other conditions. The prototypes were updated with redesigned parts and sent to West Farms near Bakersfield, California. The West Farms operated approximately 30,000 acres of dry land and irrigated crops. From January through October, we could often test the tractors for 24 hours per day.

Joe Roland was in charge of these tests. He provided a weekly report on the operation of each tractor and other equipment, such as South Bend's Rollover plow. The Rollover plow had right and left plow bottoms so that a field could be plowed by passing through the field, roll over the plow to the opposite plow bottoms and then plow back in the same furrow. This type of plow was important to maintain level fields for irrigation. West Farms was used as a testing area until about 1958 when the president of West Farms died.

The next test area was near Lubbock, Texas. The tractors were operated on the plateau where there was much dry land and some irrigated farming. About the same time the testing was moved to the Lubbock area, Alva Phelps, our Oliver CEO, had a friend in Mississippi who promised good usage of some test equipment, so some of the test tractors were sent to Mississippi. But neither the Lubbock nor the Mississippi areas were satisfactory for our testing purposes.

In 1960 we moved Joe Roland and our test equipment to the Phoenix area. We first worked with the Phoenix Vegetable Company. The test tractors and equipment were loaned to anyone in that area who could meet our test requirements. A Quonset hut was leased as headquarters for our equipment and for any required repair. This building was located a short distance south of the southeast corner of the original Sun City. I usually stayed at the Kings Inn in Sun City when I visited the test areas to observe the test operation and the equipment. I watched Sun City grow during the 1960s. This test area continued to be used as long as Oliver and White Farm Equipment Engineering had such a remote test program.

Tractor tests were also conducted all year round near Charles City or in the upper Midwest. Industrial wheel tractors were tested in our laboratory with stress coating and strain gauges. Industrial tractors were tested anywhere that we could obtain the best answers to strength, wear, and the appropriateness of the application of the tractor with its equipment.

Tests equivalent to the Nebraska tests were conducted on the Fleetline models by testing on the streets near the plant and in the dynamometer laboratory. Later, a test track was built on Oliver"s property in Charles City. The outside track was constructed to be comparable to the Nebraska test track. The inside track was an obstacle course with logs bolted to the concrete. These logs can be seen in the photo just behind the tractor and dynamometer being tested on the outer track. The purpose of the obstacle course was to predetermine which parts of the tractor were subject to early fatigue failure. Test tractors and equipment rolling over the logs at predetermined speeds caused shock load conditions.

Engine and Transmission Testing

Engines required numerous tests to provide the best combination. Before the introduction of the latest oil bath and dry type air cleaners,

The Oliver tractor test track located at Charles City, Iowa, was similar to the test track at the University of Nebraska at Lincoln. The Oliver test track also had an obstacle course where the tractors were driven over logs as shown in the background here. ***Carl Rabe***

we had a dust room where an engine could run continuously under load in a controlled dusty atmosphere. The dust was brought from Arizona. The wear areas in the engine used for the dust room test were measured before and after each test. The purpose of the test was to determine the most efficient air cleaner for each model of engine.

Engines were also tested on a dynamometer with a direct hookup or on a belt-driven dynamometer. Engine power and fuel consumption were measured for each combination of intake manifolds, air cleaners, radiator, and other such engine accessories. The purpose of this test was to establish the most practical combination to provide long life and low fuel consumption. It was not unusual for an engine to be tested for 2,000–3,000 dynamometer hours to arrive at the best combination of parts.

The transmission, differential, and final drive, including the axles, was an extremely important assembly. One test, called a four square test, used two units opposing each other. A tractor engine does its work by applying a rotating force on the transmission shaft. The four square test utilized the two transmissions, differentials, and final drives and a very large sprocket fastened to each axle with a large roller chain connecting the sprockets on each side. The transmission input shafts were rotated in opposite directions until the equivalent force of the engine was reached. The two input shafts were then fastened to provide the desired rotating force or torque. Then, it was only necessary to rotate the input shafts at the engine governed speed. The four square test was especially severe because the test torque was established at full engine power while the average load in the field only utilized about 55 percent of the available engine power.

We attempted to judge how a tractor would be used and determined an average percentage of operation for each of the six transmission speeds. We then tested the units in each speed according to the estimated percentage of time it would spend in that speed for a total of 1,000 hours. There were frequent inspections to observe and measure wear patterns. The lowest speeds were discounted to some degree because the wheels would spin out before there was enough traction to use all of the engine's power. Some of the higher speeds used the same transmission gears as other speeds so that the

Oliver's Hydra-Lectric unit, used for lifting implements, was tested in the Experimental Engineering laboratories. This cycling test subjected the valves, levers, and electrical connections to continuous up and down operation, giving the hydraulic valves a severe test. *Carl Rabe*

The Hydra-Lectric cylinder was operated at maximum hydraulic oil pressures, lifting a weight equivalent to that of the tractor. This was a severe test because the required amount of hydraulic oil pressure under normal operation was only a quarter of the maximum, and the test subjected the cylinder to rapid cycling where normal use would only require intermittent lifting. *Carl Rabe*

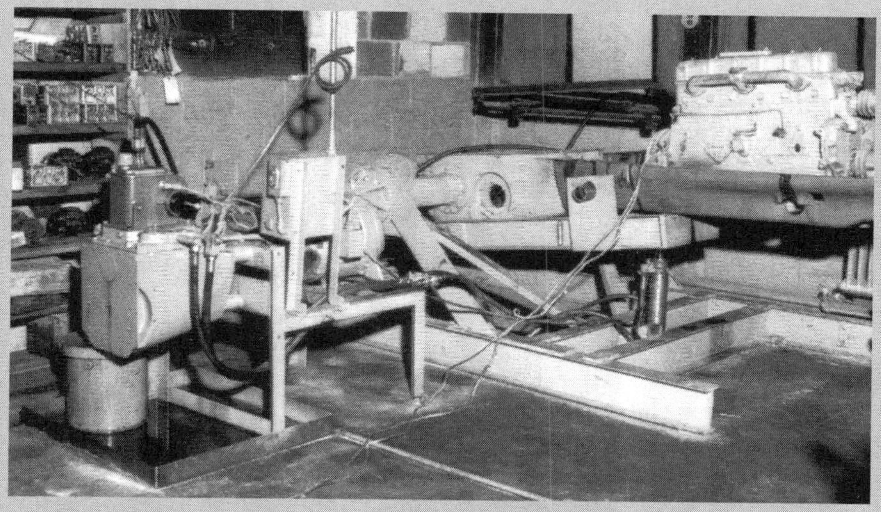

The Hydra-Lectric unit was also used to test the brake bands. The rapid cycling of the brake bands increased the temperature so much that it caused the brake lining to fail. ***Carl Rabe***

road speed was not an important factor in determining the final design of the gears.

Laboratory tests were conducted on other individual assemblies, such as the pulley, power take-off, and hydraulics. The pulley with the link attached rotated slowly. Each revolution of this pulley provided a raise and a lower cycle of the Hydra-Lectric cylinder. It was cycled to provide continuous lifting and lowering of the tractor front frame and engine. When the load was lifted, the unit would stop momentarily and then the valve would open to lower the load. This test unit was operated at near full relief valve pressure for 1,000 hours or more. Some parts were redesigned as the result of this test.

This test was quite severe compared to actual field operation. First, usage in the field required infrequent actuation of the hydraulic system. Second, average actuations required only 30–50 percent of the available pressure. The life of the hydraulic system under these normal conditions would be several times the life of a unit operated under the severe laboratory conditions. The brakes were cycled by the use of a Hydra-Lectric system to actuate the brakes frequently.

These are just a few examples of hundreds of test setups in the laboratory and in the fields. Charlie Adams, a member of our

This 1956 photograph shows the Charles City plant's inventory of recently built tractors ready for customers. ***Harold W. Snyder***

Experimental Engineering Department, had experience with the University of Nebraska's tractor test laboratories before coming to Oliver. He was an expert at setting up special test fixtures. He was slow and methodical. We had brainstorming sessions to design tests to overcome field problems. If we asked him for a commitment on when a test could start, he was vague. But with some help from us, a target date would be established. Charlie nearly always completed his part of the assignment ahead of schedule.

CHAPTER TWO

Development of the Fleetline Tractors

Design Criteria

In 1940 Oliver began research to determine all of the anticipated applications, configurations, accessories, and special tractors that would be required in the foreseeable future to establish the new tractor line, the Oliver Fleetline. These tractors were to be new from the ground up.

The Oliver Fleetline tractors represented a shift in Oliver's approach to new models. Designed with part and component group interchangeability in mind, the new line was engineered for maximum cost effectiveness. The new line also introduced a host of new features and an unprecedented emphasis on operator comfort. *Jeff Hackett*

The Oliver 60, 70, and 80 models had been quite successful, but they were designed individually without much consideration of other models from the standpoint of interchangeability of parts or component assemblies. New innovations such as an independent power take-off had not been very practical. Another consideration was to have quiet tractors in the sound range of 85 decibels or lower at the operator's ear. Overall appearance and styling, engine fuel options, maximum interchangeability of parts among the three new tractors, and the well-being of the operator were a few of the most important design criteria for the Oliver Fleetline.

Standard Tractor Configurations

The early Hart-Parr and competitive tractors were designed primarily for drawbar work—pulling plows, disk harrows and other such implements. Some tractors also had a belt pulley mechanism to power some stationary pieces of equipment, such as corn shellers, threshers, or rock crushers.

These tractors were very large and heavy with steel wheels and lugs which made them effective in pulling large implements in the field, but they were very slow.

Can you imagine a very large tractor pulling plows with many 12-in. bottoms at about 1 1/2 mph? Some stories that have come from the large wheat fields in Montana tell of farmers who plowed until noon to reach the other end of the field, ate their lunch, and spent the remainder of the day plowing to get back home. It must have seemed to take forever to plow that big field. I have also heard of a farmer who had a cabin at the one end of the field where he slept at night and then spent the next day plowing to get back home.

The many different tractor configurations needed to meet customer requirements were considered during the early stages of planning the new Oliver Fleetline tractors. The term Standard was introduced by Oliver to distinguish the regular tractor configuration used for drawbar and belt drive work from the new Row Crop tractors. The Standard tractors had a fixed tread front axle and fixed or limited rear axle tread adjustment. The front axle was low and didn't provide for much crop clearance. It was about the same height as the center of the front wheel.

The Oliver 66, 77, and 88 tractors were introduced at state fairs, local fairs, and at dealer's programs. This photo is typical of the setting for the introduction of the new line. Also available was literature highlighting the new innovations featured on the tractors. ***T. Herbert Morrell collection***

The rear tires had a smaller outside diameter than the Row Crop tractors. The Wheatland version was the Standard tractor used where there were large fields of small grain with little or no requirement to cultivate row crops.

The Orchard version was similar to the Standard except it had a large, extra sheet metal cowling to provide smooth lifting of any limbs encountered and to protect the operator. The steering wheel and operator's seat were lower and farther back, and the steering wheel was below the special cowling so that it would not interfere with tree limbs.

Some customers thought they knew a better way to design a tractor. During the late 1940s a customer named Hutchinson from Orlando, Florida, was not satisfied with Oliver's Orchard tractor design. He commented that Oliver did not know how to design a good orchard tractor. Chief engineer Louis Gilmer invited him to Charles City to tell Engineering how to design a better orchard tractor. Hutchinson was referred to our Experimental Engineering Department where parts were to be made according to his instructions.

He insisted that the operator's seat be able to pivot to different operator positions and that it be easily removed from its mounting without tools when the seat was not needed because of equipment attached to the tractor. The parts were made and installed according to Hutchinson's instructions.

Next came the trial to check out the newly made parts. It was tested on E Street, which ran through the plant. South of the plant were three sets of railroad tracks. Hutchinson drove the tractor south on E Street across the railroad tracks, turned the tractor around, opened the throttle wide open in sixth gear and drove back across the tracks at about 17 mph. The seat came loose from its support and left Hutchinson and the seat on the railroad tracks. The tractor crashed into the cars parked between the plant's buildings on E Street. The first car was totaled; the second car had considerable damage; and the third and fourth cars required some repair.

Hutchinson's parting comments were, "I am sorry about the accident. I hope that you understand what I was trying to accomplish. I leave the tractor design to you."

The Ricefield version was similar to the Wheatland version. The main difference was in the rear tires which had deeper lugs to penetrate the water and mud of rice fields to the more solid soil underneath. Moving parts that would normally be below the top of the water in the rice field were sealed.

The Industrial version was similar to the Standard except the front axle was strengthened to support heavy front end loads when implements such as front loaders were attached. The rear axle carrier was also strengthened to support rear end loads for attachments such as backhoes. The front and rear tires had higher load-carrying capacities to support the heavier loads. Low-bar tread tires were used when operating on paving or compacted soil. Diamond tread tires were used in sand and sandy soil conditions.

Row Crop Tractors

Oliver registered two patents for the Row Crop badge with a rectangular base and for the Oliver/Hart-Parr Row Crop logo on December 30, 1929. In those patents, the company stated that they have been using this logo since November 8, 1929.

Oliver's Fleetline tractors were designed with interchangeable parts and operator comfort in mind. The new line—the 66, 77, and 88—was put into regular production in 1928 and 1929. This is a 1950 Oliver Standard 66. ***Jeff Hackett***

The Row Crop tractors had larger outside-diameter rear tires than the standard tractors to provide adequate clearance for most crops. Some crops, such as sugar cane, require more crop clearance depending upon the method of growing the crop.

The tricycle Row Crop tractors came in two versions. The dual front wheel was the most popular in the 1940s and 1950s. It had two small front tires and wheels mounted close together so that they could go between two rows when cultivating crops such as corn. The single front wheel was similar to the dual front wheel version except that it had one larger tire in front instead of the smaller two tires. The single-front-wheel version was used primarily for cultivating special vegetable crops such as asparagus.

In the mid-1950s, after Ferguson and Massey Harris merged to form Massey-Ferguson, the company sent a letter threatening Oliver with a lawsuit for patent infringement on single-front-wheel tractors. Massey-Ferguson had just received a patent, but Oliver had advertised and sold single-front-wheel tractors since the 18-27 Row Crop in 1930. Copies of advertising literature and a dated assembly drawing were sent to the Massey-Ferguson legal department showing that the patent was worthless because one cannot receive a patent on something after it has been sold to the public for one year or more.

The Row Crop 77 was built from 1948 to 1954. The tractor was available with a six-cylinder gas or diesel engine. This 1949 model is distinguished by the yellow stripe on the top of the grill, a feature exclusive to 1948 and 1949 models. This tractor is equipped with the mechanical power lift and independent PTO.
Jeff Hackett

Another popular Row Crop version in the 1940s had an adjustable front axle. It became more popular later because it was more stable than the tricycle types. The front axle was adjustable for various row spacings and high enough to have about the same crop clearance as under the rear axle housings. The front tires could be aligned with the rear tires so the rear tires would make tracks within the tracks made by the front tires.

The Extra-High-Clearance Row Crop tractor had an adjustable front axle designed primarily for sugar cane fields in Louisiana where the sugar cane is planted on ridges and the tractor tires are operated in furrows. The design required approximately 12 inches more crop clearance than the regular Row Crop tractors. Drop gear housings with chain or gears as a drive mechanism were normally used to obtain the extra height. The front axle was adjustable, similar to the regular Row Crop, except that the front axle had to be about

12 inches higher. This tractor version looked like it was on stilts. The rear tires had deep lugs and the tires had a large outside diameter that made the tractor suitable for operation in the muddy furrows. Another application was for cultivating special vegetable crops such as asparagus. Because of its specialized uses, sales of this version were low compared to the other Row Crop versions.

Styling for Eye Appeal

By the mid-1930s, there was some effort to design tractors for more eye appeal. The wives on the farms began to influence the purchase of tractors, automobiles, and other equipment with eye appeal as one factor in deciding which brand to choose. In 1935 new styling on the Oliver 70 included an enclosed engine compartment. In 1938 the Oliver 70 appearance was improved again. In 1940 the Oliver 60 model was introduced with new styling and the 70 tractor was then restyled to match the 60.

The design of the new styling for the Oliver Fleetline tractors began in 1942. In 1944 six experimental prototypes of each of the Models 66, 77, and 88 were built for test and development purposes. These tractors were known as XL, XM, and XK, respectively. The styling of these experimental tractors resembled the styling of the model 60 so it was not obvious that these tractors were of a new design when they were observed in the field while being tested. Also, Oliver Engineering needed more time to complete the sheet metal and general styling for an effective future introduction to the public.

To assist with the new look, Oliver employed industrial design consultant Wilbur Henry Adams to work with all of the Oliver plants on new products. One of my first assignments in October 1944 was to work with him to design the wheel guards for the new Fleetline tractors. Wilbur would make an artist sketch of a design to be compatible with the styling of the other parts of the tractor. It was then my job to make a design layout and details of his sketch to fit the tractors.

Wilbur was primarily concerned with the eye appeal and the overall design and he had no responsibility for the tooling and product cost factors. When a design was complete, it was my responsibility to review the design with our Manufacturing Engineering Department to obtain preliminary cost estimates. Wilbur's designs were beautiful and exotic for that era. I learned very much from him.

But his first designs cost far more than we could afford. While he was away from our plant, I got permission to propose my own design, which was the fourth of our 12 designs. The cost estimates for tooling and the product on previous designs and suggestions from our Manufacturing Engineering Department influenced my new design.

My design could be made in two sizes and by having some extra tapped holes in the flange of the axle carrier, the wheel guards could be interchangeable for both the left and the right sides of the tractor. The larger wheel guard would fit the tractors having the larger-outside-diameter tires of the Row Crop and the smaller wheel guard would fit the tractors having the smaller-outside-diameter tires such as the Wheatland. Wilbur agreed with the need for some compromises in his designs.

Most tractor wheel guards had lamp brackets bolted to the wheel guard sheet metal. Within a few years, the metal cracked around the lamp bracket. I was determined that such a problem would not exist on my design. Three formed channels were designed to make the wheel guard strong and provide a solid means of attaching the wheel guard to the axle carrier. To avoid failure of the wheel guard around the lamp bracket, the attaching bolts were to go through the lamp bracket, through the wheel guard and through the channel. I am not aware of any failure of the wheel guard around the lamp bracket on any of the Fleetline tractors. My wheel guard design lasted until the Oliver 1800 and 1900 tractors were introduced in 1960. Many of the competitive tractors in the United States and foreign countries copied the important points of this design.

I also designed wheel guard extensions for the two sizes of wheel guards. These extensions were released for production as kits that could be added to the tractors before or after the tractors were sold. The extensions reached out over part of the rear tire.

I was involved with the sheet metal styling design in other areas that were my responsibility for the final design, including the transmission, differential, final drive, mechanical power lift, pulley, power take-off, seat and related parts for all three Fleetline 66, 77, and 88 tractors. I was often referred to as "the rear end man." I followed the different styling changes through to the final design.

During late 1945 and early 1946, a pilot run of 300 Model 88 (XK) tractors were manufactured for the purpose of testing the market in the field prior to public introduction. Again, the sheet metal resembled the Model 60 tractor, so the new styling could be kept a secret until the introduction of the Fleetline. Interchangeable parts and low-cost tooling were used until the final design was agreed upon by Oliver management, sales, engineering, manufacturing engineering and all others concerned. The acceptance of the pilot lot 88 tractors was great.

During 1946 the new styling was approved. The first production of the model 88s had the same sheet metal as on the pilot 300 units. In the meantime, 300 pilot tractors of the 77 and 66 were manufactured to test the market for these two sizes.

Engines

One of the most important decisions in planning Oliver's famous Fleetline tractors was the design of the engines. Determining the sizes of the engines was especially important for the full line of tractors. The final decision was to design a four-cylinder engine that would provide a net belt-pulley or power-take-off horsepower of 22 for the 66. The 77 was to be a six-cylinder engine of the same bore and stroke giving a net horsepower of 33. The 88 engine was to be a larger six-cylinder 77 engine providing a net horsepower of 44 from the belt pulley or the power take-off.

Another challenge was determining where to build the engines. There were not enough laborers available in the Charles City area to build 25,000 to 30,000 tractors per year and also manufacture the engines. Waukesha Motors Co. in Waukesha, Wisconsin, had manufactured the Oliver 60 and 80 engines, and was quite interested in a contract with Oliver to machine the castings and other parts, and assemble the engines for the new Fleetline tractors.

Charles City plant representatives worked with Waukesha's President James Delong, Vice President of Sales Frederick Schulze, and Senior Project Engineer Roger Merriam. Merriam was assigned to work full time on the Oliver account. We had great cooperation and understanding with these gentlemen so our working agreement was outstanding and long-lived.

We made an arrangement with McCoy Truck Lines using two trucks, one operating in Iowa and the other in Wisconsin. The Iowa truck transported the castings from Oliver's foundry in Charles City to Prairie du Chien, Wisconsin, where they were picked up by the Wisconsin truck and taken to the Waukesha Motors plant in Waukesha. On the return trip, the Wisconsin truck brought a trailer load of engines to Prairie du Chien where they were picked up by the Iowa truck and taken to the Oliver plant. Interstate Commerce Commission regulations and economic considerations made this arrangement advantageous.

To obtain maximum power and fuel economy from the steadily increasing octane rating—expected to be 74 when the tractors would start into production—Oliver selected a compression ratio of 6.75:1 for the gasoline engine. For the engine that burned kerosene—fuel with a 35 octane rating—a compression ratio of 4.75:1 was selected. The compression ratio of 15:1 was selected for the engine that would burn diesel—a fuel with a 40-octane rating. The rated and governed speed of these engines was 1600 rpm. The Oliver Diesels were so successful that the kerosene engines were rapidly losing appeal.

Tests at Waukesha Motor Co. determined that the peak combustion pressure for the diesel was approximately 1,100 pounds per square inch. The peak combustion pressure on the gasoline engine was approximately 850 pounds per square inch. Oliver chose to apply many of the diesel engine principles to its gasoline engine to achieve high brake mean effective pressure. The crankcase and all other parts below the cylinder head were common to the gasoline, kerosene, and LPG versions of each size engine. This design strategy produced both rugged and precision spark-ignition engines with resultant excellent fuel economy and durability.

Development of the Diesel Engine

The first 50 Diesel Model 88 tractors were manufactured with the Ricardo combustion system in the cylinder head. There was not enough clearance on the Row Crop tractors to use a Robert Bosch or an American Bosch in-line multiple-plunger fuel injection pump.

At first, neither American Bosch nor Robert Bosch was interested in designing a single-plunger fuel injection pump, so the first 50 diesel 88 tractors were assembled with an experimental single-plunger fuel injection pump manufactured by Sundstrand. The design of this single-plunger pump was a cooperative effort among Sundstrand, Waukesha, and Oliver. After these 50 diesel 88 tractors were distributed in 1948, American Bosch proposed and then produced a single plunger pump in time for the first production run of the diesel 88.

In the meantime, I was responsible for the final design of the transmission, differential, final drive, and other attachments and accessories behind the front frame of the tractor. There were tests being conducted on diesel 88s and some experimental 66s and 77s near Bakersfield, California. I reviewed the weekly performance reports on each tractor being tested and noted that at 40 degrees F, the diesels were hard to start.

My brother Ray Morrell near Blue Mound, Kansas, wanted to buy an 88 gasoline tractor, so arrangements were made between our plant and the Kansas City sales branch. Then I got a call from the Kansas City branch saying that all of the gasoline 88s had been allocated so they wanted to substitute a pilot run Diesel 88. I rejected the substitution because my brother had herds of livestock and needed a reliable tractor for operation in cold weather. He got his 88 from the next production run.

Word of the rejection traveled quickly from the Kansas City branch to Oliver's Chicago office and then to Charles City. "Who is this engineer that doesn't want his brother to have a diesel 88?" they asked. After that, the cylinder head fuel combustion system was changed from the Ricardo system to the Lanova system with the American Bosch single-plunger fuel injection pump and the electric air intake heater provided a very good cold weather starting diesel. In fact, some farmers said they started Oliver Diesel tractors in cold weather and then pulled their gasoline automobiles to get them started.

Diesel Difficulties

Were there any problems? Yes! Because of the short time for development of the American Bosch single-plunger pump, the control linkage was not substantial enough for long life. When the linkage failed, the fuel meter fell down and let the engine run over the rated speed because of all of the extra fuel. Fortunately, the top engine speed was limited by the size of the air intake manifold.

As soon as the problem was discovered, our Charles City Engineering Department was in touch with American Bosch to provide a better linkage. At Charles City we designed a test that was severe enough to fail a production control within a few hours. Between our rigorous tests and the cooperative effort with American Bosch, a satisfactory control was provided. We replaced the controls on the tractors that had already been manufactured. I am not aware of any failures of the new control linkage.

Only a few control linkages had failed before this problem was solved, but we heard some interesting reactions to this overspeeding situation. One farmer thought an airplane was about to crash near him so he jumped off the tractor and hid by the rear wheel. He finally turned off the fuel supply to stop the engine. He then contacted his dealer for some service. The control was replaced and the farmer was back in operation again.

Another important part of the diesel development was the Roosa fuel injection pump invented by Vernon Roosa. The Roosa pump was simpler and cost less than the single-plunger Bosch fuel injection pump. The initial dynamometer tests in Oliver's Experimental Engineering were very encouraging. Oliver continued to test the Roosa pump for several months and make recommendations for improvements. The pump was also installed on some of the test tractors in the field. The Roosa fuel injection pump soon became Oliver's choice for production.

Selling the Oliver Diesel

We had another problem at the beginning of our diesel program. Farmers were not eager to buy a diesel tractor because of the poor reputation of the International Harvester Super M diesel which had been very unreliable. To provide farmers with an incentive to buy Oliver diesel tractors, Oliver started a new advertising program, "Buy an Oliver Diesel and Oliver will send you a check for half of your first six months' fuel bill." This program sold a large number of diesels and the cost was an average of $44 per tractor. Oliver tractor owners were telling their neighbors about the fuel savings compared to the gasoline tractors. In 1954, before the competition caught up to Oliver, Oliver was selling 45–50 percent of the diesel agricultural tractors.

Engine Lubrication

Oliver patented a low-pressure spurt-type engine oil lubricating system, which forced lubricating oil through the crankshaft at a pressure of about 15 psi. A slot in the crankshaft registered with a slot in the connecting rod of each cylinder. When the two slots were registered, lubricating oil was forced into the connecting rod. A limited amount of oil reached the cylinder wall from each connecting rod and the oil consumption was decreased. However, this type of system was not practical on tractors having full-load speeds above 1,600 rpm. The combination of speed and size of the slots caused concern at higher speeds.

Oliver at Seabrook Farms

Seabrook Farms, Seabrook, New Jersey, grew vegetables of nearly all types for Birds Eye, a company that sells frozen vegetables. Seabrook Farms started during the Depression and was quite successful. They leased about 30,000 acres and contracted with other farmers who grew vegetables on another 50,000 acres.

By 1949 Seabrook Farms were ready to replace their tractors. Some of the contracted vegetables were of higher quality than those grown by their company and research revealed that the farmers were all using the same fertilizer and general crop management as Seabrook Farms. The best vegetables were grown by farmers who had Oliver tractors. They commented that the tractor and cultivator combination provided more accurate control. As one farmer explained it, “You can put the fertilizer and the cultivator shovels right where you want them.” This survey, along with the Oliver diesel program with independent power take-off and the many other advantages, were major factors that led Seabrook Farms to begin replacing their tractors with Oliver diesels. In March 1950 we conducted sales and service schools for Seabrook Farms employees to familiarize them with the 69 Oliver diesel 66, 77, and 88 tractors that they had purchased. This was the start of a program to replace approximately 300 tractors.

I was called back to Seabrook Farms in September to resolve a combination lubricating oil and fuel problem. They reported that one 88 diesel was using six quarts of oil per day and others had similar problems. We discovered that the lubricating oil supplied to Seabrook Farms by an Amoco dealer in Shiloh, New Jersey, was for gasoline engines and the fuel was a mixture of kerosene and furnace oil. Furnace oil had up to three times more sulfur than diesel fuel.

Excessive sulfur in the fuel is bad news for diesel tractors, particularly for those operating at or near full load, and the Seabrook tractors having the problems were operating at full load.

We examined the tractor reported to use six quarts of crankcase oil per day. The engine was so full of sludge that one could get no more than three-quarters quart of oil into it. The tractor operator had reported that he had to fill it with oil each day. Someone in the office had referred to the operator's manual and, noting that the capacity was six quarts of oil, had assumed the operator was adding that much each day.

We removed the oil pan and the valve cover. Both were so full of sludge that there were only grooves for the moving parts of the engine. We carefully removed what sludge that we could by hand, ordered a barrel of Diesel flushing oil, changed the two large lubricating oil filters, and filled the crankcase with flushing oil. The tractor was operated at about half of the full power for four hours. The valve cover and oil pan were then removed and most of the sludge inside the engine had been removed. We repeated the four-hour cleaning operation and then we inspected the inside of the engine. It was clean and just like new.

The chief of the Amoco Research Center in Baltimore met with me and representatives of Seabrook Farms. We reviewed the problems and requested that Amoco immediately send a truck to each of Seabrook Farms Divisions to deliver heavy-duty diesel lubricating oil and remove the lubricating oil for gasoline engines, and replace the fuel mixture with No. 2 diesel fuel. Seabrook Farms and I assured Amoco that we would recommend another source for diesel products if they did not comply with our request. Amoco trucks were at Seabrook Farms the following Monday.

There were no more fuel or lubricant problems at Seabrook Farms. But this story has a rather sad ending because 1951, 1952, and 1953 were extreme drought years in New Jersey. Seabrook Farms went into bankruptcy and we did not replace any more old tractors.

Oil Seals

Oil seals were not expensive if correctly selected and applied, but if an oil seal failed, it could be very costly. Chicago Rawhide Co. was one of the most helpful and dependable suppliers. Jack Eirk was their sales representative calling on Oliver. National Oil Seal Co. was another source for good oil seals.

Independent Power Take-Off

There is a myth that Cockshutt of Canada was the first tractor with independent power take-off (PTO) in 1948. Actually, this is wrong for several reasons. First, the Hart-Parr 18-36 manufactured in 1928 had an independent PTO. Second, Oliver had a pilot run of 300 88s with independent PTO in customer's hands by early 1946. Oliver did not publicize the independent PTO until 1948, when it introduced the Fleetline 66, 77, and 88 with the new styling and other innovations. From the late 1930s to 1948, Oliver manufactured tractors for Cockshutt using the Oliver 60, 70, and 80 tractors with some different sheet metal, color and decal treatment, but these were not offered with independent PTO.

Oliver's independent PTO established a trend, at least on the larger tractors. A transmission-driven PTO was satisfactory for some implements, such as mowers. But implements, such as pull combines, could easily become clogged and the tractor would have to be stopped which also stopped the clogged machine. If the clogged machine was only partly clogged, the tractor transmission could be shifted into neutral, and the tractor power used to clear the machine. Otherwise, these clogged machines had to be cleared by the hands of the operator.

This prompted manufactures to install a separate engine on their implements costing an additional $1,000–$2,000 in the early

1950s. Oliver introduced its independent PTO for a list price of about $125, which allowed one tractor to be used for more than one implement. The introduction of the independent PTO became a great cost saving to Oliver's farmer customers.

Development of the Independent PTO

My second main assignment at Oliver was to work with a project engineer, Don Kinch, to develop a rear mounted combination belt pulley and power take-off (PTO) for the new Fleetline 66, 77, and 88 tractors. Power take-off means that power from the tractor engine is transmitted to an implement so that it can perform its function by power from the tractor engine rather than by a ground drive or some other means. My main contribution to this design

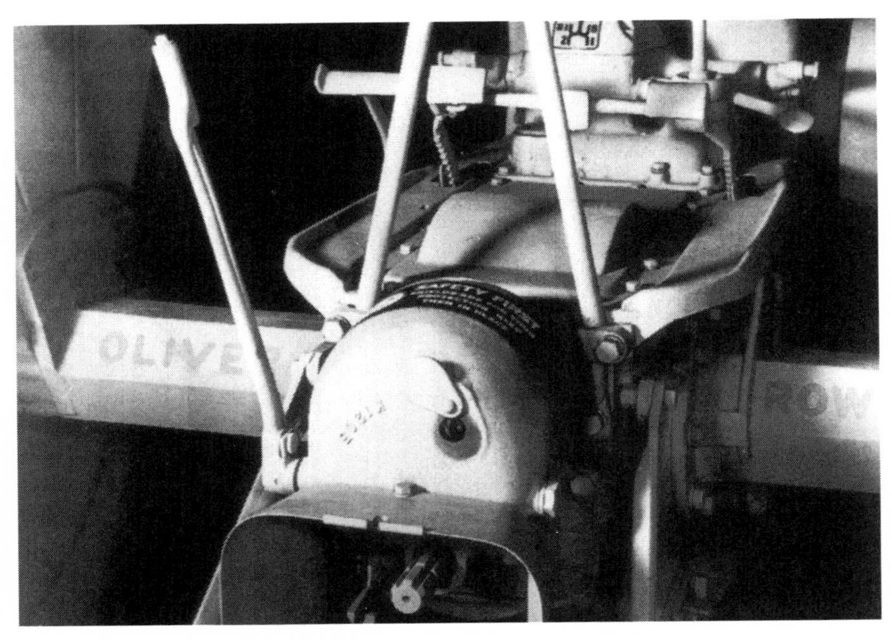

Independent power take-off (PTO) was one of Oliver's most important innovations. Although it was initially introduced by Hart-Parr in 1928, the advantages were not clear at that time. Oliver offered a practical independent PTO on the new Fleetline tractors. The model 88 pilot run tractors and first production 77s and 88s until January 1949 were made with the PTO lever on the left side. The mechanical power lift, clutch pedal, and equalizer brake pedals are also shown here. ***Carl Rabe***

was the control system, including the engaging lever and internal parts for engaging and disengaging the PTO clutch independent of the tractor main clutch.

The usual accelerated test was set up in the Experimental Engineering Department. The test continued at full load, which was equivalent to the engine power of the 88 tractor, while the clutch was engaged and disengaged frequently. If the unit was operated under these extreme conditions for 100 hours or more without failure or problems, it would be considered satisfactory for a pilot lot production and further tests. This unit failed early in the test. The clutch failed, the unit overheated, and there were other problems. It was then April 1945 and a new design had to be accomplished quickly.

I was asked to give my evaluation during a meeting with Chief Engineer Louis Gilmer, Assistant Chief Engineer Milford Stewart, and Supervisor of Experimental Engineering Tom Martin. We agreed that there was not enough room at the rear of the tractor for a combination PTO and belt pulley when considering other future designs. Besides, we predicted that the PTO would soon replace the belt pulley.

There were too many cost compromises in the combination unit, so it was suggested that the belt pulley be a separate unit mounted in front of the transmission with the pulley on the right side of the tractor. I was asked to make a preliminary belt pulley design to determine if it would be feasible to locate the pulley in front of the transmission. Such a design appeared to be quite satisfactory.

The independent PTO then took priority over the belt pulley. Don Kinch, an expert mathematician, was assigned to other parts of the Fleetline design, so I was asked to be responsible for the design of the PTO and belt pulley units for the Fleetline 66, 77, and 88 tractors. The design criteria were reviewed and timetables established. September 1945 was the deadline for completing the design and experimental tests so we would be ready for field tests. I was told that the design had to be good because we planned to release it for production and make casting patterns, fixtures, and tools during the field tests.

The main design goal was to provide an independent PTO with its own clutch so that it could be engaged and disengaged regardless of the motion of the tractor. This was accomplished by using a tubular clutch shaft and transmission input shaft. The drive shaft to the PTO clutch was splined to the engine flywheel, and passed through the clutch and transmission shafts to an independent clutch at the rear of the tractor.

The pilot production plan was to manufacture 300 88s with the new independent PTO during late 1945 and early 1946. By May 1, I had the PTO design ready for the drafting department to make drawings of all the individual parts. Oliver's policy was that the designer would not make production drawings of his design or check them. The drawings were to be made and checked by someone other than the designer. However, the drafting department was so busy with the drawings of other parts of the tractors that they did not know when they could start on my design, so I made the production drawings of all the parts. When I asked to get them checked, no design checker was available, so I checked the drawings and our specifications department released the drawings for experimental parts. We ordered three sets of parts, one for laboratory tests, one to be installed on an experimental 88 tractor for field tests, and the third set for repairs or a second field test unit.

The accelerated test on the experimental PTO was completed with only minor issues which were easily corrected. The field test was done through a farmer who had a large acreage of crops and was a good customer of Thill Brothers Implement in Rose Creek, Minnesota. This farmer became a custom operator using the experimental 88 with a pull silage harvester for filling silos with corn stalks. Joe Roland was the field test engineer following these PTO field tests. These tests were also used to prove the experimental Model 88 tractor in general. Since there were good results from the tests in the laboratory and in the field, it was then full speed ahead to release the Model 88 tractor and PTO for the pilot run.

After the pilot run, we encountered some problems with the independent PTO system at first. The long drive shaft between the engine flywheel and the PTO clutch acted as a torsion spring to damp shock loads. We first chose Stressproof steel for the long drive shaft because it was advertised to be very strong and wear resistant, but advertising can be misleading. One of my first field trips for

Oliver was in late spring 1946 to Herman Streakers Implement dealership in Wapakoneta, Ohio, to observe some problems with a pilot run Model 88 tractor. After the customer had used his PTO a considerable number of hours, the splines were worn off the drive shaft at the flywheel end and one bull gear in the final drive was worn considerably. It was obvious that the bull gear had been carburized but missed the hardening operation.

Because some late tests had indicated that the PTO drive shaft needed to be made from stronger, highly heat-treated steel to avoid wear, we were experimenting with 4140 steel. We used electrical induction to harden the splined areas and the smaller part of the drive shaft that passed through the tubular clutch and transmission shafts. I called the plant at Charles City and asked that a new induction-hardened drive shaft, a new induction-hardened flywheel hub with splines and a new bull gear be shipped as quickly as possible. These were sent at no charge to the dealer or the customer and Oliver reimbursed the dealer for his mechanic's time to install the new parts. Oliver established a service program to replace all the PTO drive shafts and flywheel hubs in the pilot run Model 88s.

If the operator engaged the independent PTO clutch quickly with the throttle wide open, and there was a large inertia required to accelerate the implement, there could be some extreme shock loads on the PTO. With the old transmission-driven PTO the operator would feel the shock through his body. My solution was to mount the PTO clutch lever on the left side of the operator's seat and provide operation guidelines. The operators were advised to reduce engine speed, engage PTO clutch slowly to accelerate the implement, and then increase the engine speed. This way both hands could be used to coordinate the PTO engagement without undue shock loads on the shafts and mechanism.

The pilot run Model 88s all had the PTO lever on the left side, but because of a deluge of requests, the lever was moved to the right side. There have been some failures of either the drive shaft or the standard output shaft which I attribute to extra shock loads from having the control lever on the right side. Such a location prompted the operator to increase the engine speed to maximum with the right hand and then quickly engage the PTO with the same right hand.

We have measured shock loads equal to ten times the continuous output of the 88 engine.

The next step was to experiment with various PTO clutch spring pressures. The high shock load occurred only for a fraction of a second until the clutch could break loose and slip.

We established our clutch spring pressure so that its capacity was only 1.25 times the maximum power of a Model 88 engine. The trade-off with that design factor was that we anticipated more PTO clutch plate failures. However, the clutch problems were not too great considering the large number of tractors and all of the available implements. Also, as time went on, the operators became more aware of the need for smooth engagement of the PTO clutch.

The Oliver independent PTO became the envy of most tractor manufacturers. Within a few years, most agricultural tractor companies in the world had designed implements for the independent PTO, at least on the larger tractors.

I represented Oliver's tractor division in the Farm and Industrial Equipment Institute (FIEI). After an FIEI meeting of the Power Take-Off Subcommittee of the Advisory Engineering Committee in 1954, Ed Tanquary, staff engineer for International Harvester

The initial design with the PTO lever on the left was not accepted by operators, so the location of the lever was changed to the right side, as shown on this 1949 Row Crop model. *Jeff Hackett*

(IHC) and chairman of the FIEI Advisory Engineering Committee, and I continued to discuss the day's activities and where we were going on PTO standards.

I asked him, "Is IHC concerned about the independent PTO or do you think Oliver is too small to be a factor in the total tractor sales?"

He seemed to be startled by my question and said, "You will never know how much Oliver's independent PTO concerns our company."

Then I said, "Oliver has had the independent PTO for six years. Why doesn't IHC have it yet?"

His answer was, "We are so big and there are so many levels of management that by the time a proposal reaches the top and gets back to the responsible design people, it is often obsolete and we have to start all over again. We are envious of you. You can prepare a proposal, send it to your Chicago office, and within a few days you have the approval to go ahead with the proposal. If IHC had your tractors with our sales department and Oliver had our tractors with your sales department, we would do extremely well and you would soon go broke. Your sales department is expecting you to keep providing something new so that the tractors will sell themselves. Your sales department has forgotten how to hard sell farm equipment."

I wrote a paper on the development of the PTO for agricultural equipment titled "The Development of Agricultural Equipment Power Take-Off Mechanism" and presented it in Milwaukee, Wisconsin, at the Society of Automotive Engineers (SAE) Farm and Construction Machinery Technical Meeting September, 1980. This paper was reprinted in the SP-470 book published in 1981 for the celebration of SAE's 75th anniversary.

A few product liability lawsuits were filed against Oliver tractors involving the power take-off (PTO). The following case is related without names. This case involved a 77 Standard tractor in South Dakota that was manufactured in 1950 and was nearly 18 years old when the accident occurred March 4, 1968.

The owner filled a DuAll silage wagon full from his silo. He then drove the tractor and wagon to his livestock feed bunks. The wagon was unloaded by engaging the PTO. The wagon mechanism moved the silage forward to a beater, a rotating mechanism with large spikes. The beater forced the silage through an opening onto a cross conveyor which then deposited the silage into the livestock feed bunk. When the owner emptied the load of silage into the bunks, he noticed some clumps of frozen silage lodged in the wagon. He disengaged the PTO and climbed into the wagon to dislodge the frozen clumps of silage. He claimed that the tractor PTO engaged itself and drew him into the beater where he suffered severe injury to one leg and some injury to the other leg. Other claims in the case were that the PTO control mechanism was not fail-safe and that there should have been an automatic safety release to prevent injury.

Oliver purchased a used 77 Standard from the same area in South Dakota that was the same age as the one involved in the accident. In fact, they were only a few serial numbers apart. The attorneys managed to buy the DuAll wagon that was involved in the accident. The tractor and the silage wagon were taken to Naperville, Illinois, for testing and evaluation.

The main claim by the plaintiff was that the PTO control system was not fail-safe. To prevent accidental clutch engagement, the PTO clutch lever did have an automatic lockout when the PTO clutch was disengaged. The pawl on the lever engaged an adjustable stop on the clutch cover. The PTO had an over-center clutch with heavy spring pressure behind it. When the PTO clutch was disengaged, the control lever snapped rearward and its pawl locked up against its adjustable stop. To engage the PTO, the operator had to push down on a gearshift-type knob at the top of the control lever to lower the pawl below its stop and then push the lever forward past the over-center position. The clutch then stay engaged until the lever was pulled rearward to disengage the clutch.

The Model 77 and 88 PTO clutch had double disc-friction plates with a center plate between them. Three small separating springs were held between the front and rear pressure plates of the clutch. These three springs passed through holes in lugs of the center plate to separate the clutch plates and the center plate. The separation of the discs and the center plate was important to prevent any drag or rotation of the PTO output shaft when the clutch was disengaged.

I was among the expert witnesses who went to South Dakota to view the tractor involved in the accident. We obtained permission to disassemble the PTO clutch and photograph the parts.

We discovered that the three separating springs were missing and the holes in the center plate and the front and rear pressure plates were filled with grease and particles from wearing of the clutch face. Also, the clutch control lever was adjusted very tightly. Possibly this was an attempt by the owner to disengage the PTO and prevent any drag by the clutch parts. This led us to assume that the owner had done his own service repair on his 77 Standard PTO. The result was that the clutch was modified and a condition was created that would permit some drag of the clutch parts when the clutch was disengaged.

In tests conducted on the tractor and the silage wagon at Naperville we learned that when the three separating springs were omitted, approximately one-quarter horsepower drag could be obtained by the clutch parts. This drag was enough to cause serious damage to a board fed into the silage wagon's beater. Fifteen horsepower was required to unload a load of silage from the DuAll wagon—approximately 60 times the power that would cause injury to a human being, so an automatic shut off would not have been practical.

We toured all of the farm buildings while we were at the owner's farm. In one single-car garage, there were many modified Oliver PTO parts. Some of these included sintered metal clutch plate facings. Sintered metal clutch facings were not practical for this application because the coefficient of friction was much lower than the non-metallic production and repair friction plates. We also noted some worn PTO clutch release bearings. These had apparently failed because of the owner's attempt to use the release bearing to prevent clutch drag after the separating springs were removed.

Another concern of ours was where the owner got the tractor. We contacted the Oliver dealer who worked with this owner and got permission to review his sales and service records with this customer. The previous dealer had died about two years before the accident, but the appropriate records had been transferred to the new dealership.

We learned that the owner had previously purchased as many as six Oliver tractors from the local dealership, the latest of which was an 88 tractor which had the same PTO except for a longer drive shaft. While there were service invoices for PTO repair on this 88 and repairs on all of the other tractors, there were none for the 77 Standard. Why were there sales and service repair invoices on all of the other Oliver tractors but none relating to the 77 Standard?

We checked sales records at the Charles City plant. This tractor, when new, was shipped to the Kansas City sales branch. However, a flood had destroyed the Kansas City records. The accident had occurred in South Dakota under Oliver's Minneapolis sales branch, which had complete records of all sales even back to the Hart/Parr days. This tractor was not sold through the Minneapolis sales branch. How did the tractor get from the Kansas City sales branch to South Dakota without being recorded in the Minneapolis branch? These questions might have been answered if the case had been tried in court, but it was settled out of court for a sum of money that should take care of the injured man for the rest of his life.

In preparation for the trial I was assigned to locate and visit at least 25 model 77 and 88 tractors within a 50-mile radius of Chicago. The purpose was to evaluate the condition of the PTO control mechanism from many standpoints, such as adjustment and clutch drag. I called all the Oliver dealers to find out if they had 77 and 88 tractors in their inventory. I was amazed to find that only two tractors were in dealers' inventories and one was to be delivered to the buyer the next day. I asked what happened to the used 77 and 88 tractors. The general answer was that farmers don't trade them. They keep them for their second tractor or put a loader on it and use it for chores. The few that are traded are sold quickly from the used tractor inventory.

I surveyed 42 Oliver 77 and 88 tractors, all of which were 15-21 years old. I found only four tractors with PTO control mechanisms in poor condition. Repair parts were sent to the dealers for these tractors. Fifteen tractors were rated "fair" because they needed adjustment of the pawl to the stop, the hand knob on top of the lever was missing, or there were other minor maintenance needs. Many of these tractors were adjusted at the time of my visit. The other 23 tractors had good maintenance and were in good operating condition. When one considers the age of these tractors and their general usage, I felt that the PTO control mechanism had proven to be very good.

The visits to these 77 and 88 tractors gave me an opportunity to visit with the customer and ask some questions. Why did they keep these old Oliver tractors? The most common answer was because of the independent PTO. They also mentioned the smooth running six-cylinder engines, easy ride seat, and the automatic Hydra-Lectric hydraulic system.

Belt Pulley Design

I designed the belt pulley when I had some spare time during the PTO project. One design criterion was that it be a special unit that could be added or removed easily, so I designed a pulley that could be removed from the flanged shaft by four cap screws from the outside, leaving no projection outside of the sheet metal outline. The belt pulley housing was to be mounted on the front of the rear frame. Some changes to the main frame foundry casting patterns were required to add bosses for the belt pulley housing mounting. An oil trough was added to feed oil to and from the transmission for adequate lubrication of the pulley mechanism and for cooling the oil.

The model 77 and 88 had a provision for a multiple reverse transmission to provide four speeds forward and four speeds in

The belt pulley was located on the right side of all three Fleetline tractors, the model 66, 77, and 88. This shows the belt pulley on a 1950 Standard 66. ***Jeff Hackett***

reverse. This option was used primarily for the northwest U.S. where pea harvesters were mounted on the tractors. The pulley housing had to be designed so that it would not interfere with a special shaft bracket or a cover when there was no multiple reverse idler shaft. But this option was short-lived as the special industry in that area developed larger and higher-capacity self-propelled pea harvesters.

Overall, the belt pulley was very successful with little or no serious service problems. Since that time, most of the belt-driven equipment, such as threshers, has been replaced by self-propelled or PTO-driven machinery.

Hydra-Lectric Implement Power Lifts

A mechanical implement power lift was developed for the first Hart-Parr and Oliver Row Crop tractors. It was controlled by a foot pedal located on the right side of the unit. One actuation of the control pedal resulted in a one-half revolution of the cross shaft, which had an arm attached to each end and provided for both left and right lift and lower of the implement. If the implement was in its working position, one actuation of the pedal would raise the implement to its non-working, or transport position. A second actuation of the control pedal rotated the shaft another revolution, which returned the implement to its working position.

The growth in the agricultural equipment industry during the 1930s and early 1940s prompted the tractor industry to consider lifting and lowering of implements by hydraulic power. John Deere was one of the first companies to introduce hydraulic power lifts. On their units the working depth of towed or mounted equipment could not be changed easily from the tractor seat when the automatic shut-off at each end of the cylinder stroke was pre-set. So, to avoid getting down each time, one had to forego the automatic shut-off and manually set the working depth each time the implement was lowered into its working position.

Oliver's design criteria included an automatic shut-off of the hydraulic cylinder lift and lower functions, and a simple means of changing the working depth from the operator's seat while the tractor and implement continued in motion. When the hydraulic control lever was moved to the implement lift position, it was held by

The new Oliver Hydra-Lectric hydraulic lift was a major innovation of the Fleetline series. It allowed an operator to change the working depth at which his implements were operating without stopping to get off the tractor. Here is a Hydra-Lectric hydraulic lift on a 1950 Row Crop 77. *Jeff Hackett*

a detent until the hydraulic cylinder reached the extended position, and then the control lever would automatically return to neutral. Also, when the control lever was moved to the implement lower position, the control lever would be held by a detent until the pre-set implement working depth was reached. Then a valve on the cylinder cut off the flow of oil to the hydraulic cylinder and the control lever would then automatically return to neutral.

The Oliver Hydra-Lectric was controlled by three-way electrical circuits. Changing the working depth on the go was easy to do. This was accomplished by moving the hydraulic control lever sideways to energize the magnet on the Hydra-Lectric hydraulic cylinder. When the control lever was moved sideways and then in the implement lower direction, the cylinder collar was pushed on the cylinder rod to change its pre-set position which resulted in increased working depth of the implement. If the operator wanted to decrease the implement working depth, the control lever would be moved sideways and in

Early Fleetline tractors such as this 1949 Row Crop 77 used this mechanical power lift, a standard piece of equipment that was designed for Hart-Parr and Oliver tractors with front-mounted equipment. A hydraulic lift became available as an option in 1950. ***Jeff Hackett***

the implement lift direction. The electric magnet held the collar and slid it on the cylinder rod until a new pre-set working depth was established. When the implement was raised to the transport position, the Hydra-Lectric cylinder was completely extended to its maximum length.

The operator didn't have to set the implement working depth by observing the cylinder and manually controlling the working depth each time after the implement was raised because the implement always returned to the pre-set position for the desired working depth. The operator could change the working depth by holding onto the control lever and nudging to change the position of the collar on the cylinder rod. The Oliver Hydra-Lectric was the most automatic and versatile hydraulic system on agricultural equipment.

The first electrical switches for the Hydra-Lectric were small toggle levers mounted on the steering column. We attempted to seal the switches to protect the electrical contact points from the ambient weather elements and prevent corrosion that would cause a malfunction of the switches. But temperature changes, rain, fertilizer, and other factors caused the seals to break down and the switches to malfunction early in the warranty period. The solution was to put the

The Hydra-Lectric hydraulic cylinder was mounted on the implement. Early versions were controlled by the two switches mounted to the left-hand side of the steering column. *Jeff Hackett*

Hydra-Lectric unit below the instrument panel with the electrical switches inside the oil reservoir where they were covered with oil. The oil protected the switch contact points and prevented them from corroding and malfunctioning.

Electrical cord ends corroded and malfunctioned unless they were properly maintained. The electrical cord between the unit and the breakaway hydraulic coupling area, and the cord to the Hydra-Lectric hydraulic cylinder, had specially molded rubber ends to protect the electrical cord male and female parts from the weather. One end had a circular groove and the mating part had an internal circular ring that fit into the groove. But this was not a perfect solution. Much study has been devoted to these corrosion problems by companies making lawn care and skid steer loader products, but I am not aware of any design that provides long-term success for sealing against all ambient and weather elements.

During the patent search, we learned that Bendix Aviation Division in North Hollywood, California, had a patent on a hydraulic cylinder controlled by electrical circuits. So Oliver worked with Bendix to manufacture our first Hydra-Lectric hydraulic cylinders, but there were quality problems in the production by Bendix.

Oliver's warranty cost was too great and the Hydra-Lectric system was poorly accepted.

Oliver Engineering suggested many changes to improve the quality, but to effectively make a change at Bendix took as long as a year or more. In March 1951 we discontinued the manufacture of Oliver's hydraulic cylinder by Bendix. Oliver Engineering made its own detail drawings making changes to improve the quality.

Some of Oliver's industrial engineering and manufacturing supervisors were skeptical that Oliver could machine to such close tolerances. But Plant Manager George Bird said, "We are going to manufacture our own hydraulic cylinders. It is up to you fellows to get the job done."

Oliver's manufacture of its own Hydra-Lectric cylinder solved many problems. Oliver received patent No. 2,707,867 on the system. This patent was issued in 1955 to Charles A. L. Ruhl and assigned to Oliver. It was one of the most complicated patents on hydraulics.

But there were still corrosion problems. Customers who did not maintain the unit properly were very critical of it. Also, competitors, who were prevented from using the same system by Oliver's patent, were very vocal about the problems of corrosion. These factors caused Oliver to offer an optional hydraulic system with manual control of implement working depths comparable to competitors' hydraulic systems. Oliver's Hydra-Lectric hydraulic system was still used, however, and rated excellent by those customers who maintained their units well.

Transmission and Final Drive

The Fleetline 66, 77, and 88 had basic four-speed (three forward and one reverse) transmissions. The input shaft and counter shaft system provided for doubling the available speeds to six forward and two in reverse. The original design of the differential spiral bevel gear and pinion provided for some flexibility, which allowed the differential to support all of the speeds.

Row Crop tractors were to have a top speed of about 12 mph. Since the Row Crop tractors had a high center of gravity, safe execution of turns was a major consideration. On two occasions, our experimental tractor test drivers turned sharply at road speed and the tricycle-type tractors rolled over. Fortunately, both drivers were thrown clear of the tractor and were not hurt seriously. So Oliver maintained a limit on the maximum speed in sixth gear, the road speed. The adjustable-front-axle Row Crop tractors were somewhat more stable.

The Standard tractor had a low center of gravity and limited oscillation of the front axle. There was no Standard tricycle model. These conditions allowed for higher road speed. The combination of the differential spiral bevel gear and pinion and the flexibility of the transmission permitted good working speeds in the road (sixth) gear of about 18 mph.

An interesting problem developed between the experimental model 88 tractors and the pilot run of 300 tractors. Some tractors would not stay in fifth gear when the tractor had a load requiring most of the tractor's power. The experimental tractors had very tight-fitting splines of the small pinion on the spiral bevel pinion transmission output shaft. These pinions were made on a production basis for the pilot run, and the heat treatment operation caused some of the pilot run pinions to enlarge in the spline area. I went to Le Mars, Iowa, the site of the test, and exchanged the fifth-gear pinion with a tight-fitting pinion on the output shaft.

Experimental Engineering set up a transmission with an opening on the top to view the transmission gears. A load was applied to the axle. One suggestion to keep the transmission in gear was to provide a heavier detent to hold the shifter rod from moving out of gear, but we could not keep the fifth-gear pinion from moving quickly out of mesh with the large countershaft gear even with a 6-foot bar. The shifter fork got so hot from the friction between the shifter fork and the shifter groove of the pinion that smoke appeared. I made a force diagram and some calculations. The fifth-gear pinion was removed and the Tool Room ground the spline from the pinion leaving only the splines under the large countershaft gear teeth. The revised fifth-gear pinion with less spline length was installed in the test transmission. This solved the problem; the fifth-gear pinion could not be pried out of mesh with the 6-foot bar. Oliver's service department created a replacement program for any tractors with the fifth-gear problem.

The experience with the fifth gear prompted a review of all the transmission gears on the 66, 77, and 88. We discovered that the double sliding gear on the input shaft could cause problems similar to the fifth-gear problem. To avoid the same problem, the involute splines under the shifter fork groove were eliminated. The splines and the gear mesh were then compatible to avoid any tendency for the gear to come out of mesh. These changes were then made to all of the comparable gears in the 66, 77, and 88 tractor transmissions.

We also designed an interlock with welded steel parts on the experimental tractors to prevent shifting two gear-shift rods at once, which would cause a gear lock-up. Still the gears did lock up under some conditions. This actually occurred with some of the experimental tractors in the Phoenix area. To show the movement of the gearshift lever, the interlock and the three gearshift rods, we laid out a 20:1 oversized scale drawing. From this layout, we decided to make steel stampings and locate the locking lugs accurately on the stamping. We used this design for the production run of the Fleetline tractors. I am not aware of any gearshift lock-up problems on the Fleetline production tractors.

Another feature on the model 77 and 88 tractors was the multiple reverse transmission, which allowed special equipment to be mounted on the rear of the tractor and the tractor to be driven in reverse. This special equipment was primarily for vegetable harvesters such as peas. An idler gear and shaft were added to the front of the transmission and four speeds forward and four speeds in reverse were available. This provided three working gears and a reasonable road speed in each direction of travel. About the same time as we were introducing the Fleetline tractors, local manufacturers in the Pacific Northwest were designing larger self-propelled vegetable harvesters, so Oliver never sold a large quantity of tractors with the multiple reverse transmission.

The Thomas Varidraulic Drive was another option on the Fleetline tractors. This drive mechanism had a fluid coupling with variable space between the stator and the rotor that was controlled by the operator. Slow ground speed could be accomplished by increasing the distance between the stator and the rotor of the fluid coupling. The maximum speed was reached by reducing the distance between the stator and rotor to the fluid coupling's minimum allowable setting. This fluid coupling was very inefficient and slow speeds were difficult to control. I recall driving a tractor with this variable fluid coupling. I started with the brakes set lightly to simulate a small drawbar pull requirement. I moved the control slowly toward the maximum speed. Suddenly, the tractor lurched forward. This was quite impractical. I do not recall that we sold any tractors with the Thomas Varidraulic Drive. If any were sold, the application had to be very special.

Sound Reduction

One of Oliver's goals was to provide quiet-running Fleetline 66, 77, and 88 tractors with a maximum loudness of 85 decibels at the operator's ear. This goal was accomplished on the experimental tractors. The transmission parts were made with minimum tooling. Some of the parts had to be honed to fit their mating parts such as gears to their shafts. The transmissions were of the spur gear type, which means that the gear teeth were parallel to the shaft. The result was very quiet transmissions on all three experimental model prototypes. Engine mufflers were selected to maximize quietness.

The pilot runs of 300 for each model tractor made greater transmission noise than the experimental prototypes. The gears were looser on the shafts and not as accurately machined. For the production tractors, we selected a muffler for the engine that was louder to place the gear noise more in the background. Helical gear transmissions were originally considered, but the cost was much greater.

When we started research on quieter transmissions, we wanted to determine the source of the noise and the frequencies (vibrations per second) that made up the overall noise. One Saturday, when the plant was closed and quiet, Tony Obermeier and I conducted the sound test. I had a C-Melody saxophone and could relate the frequencies by harmonizing with the transmission sound. Tony ran the tractor on the inspection treadmill while I scanned the notes on the saxophone. We thought that the predominate frequency would be the same as the number of teeth in mesh per second between the input gear and its mating gear on the counter-shaft. I found it easy to harmonize with the predominate frequency. The remaining frequencies were generated by the engine, other gear combinations, and other functions of the tractor.

To continue our research, we sought a consultant that was experienced in noise reduction of commercial products. We hired Silencing Consultants of America in Toledo, Ohio, who had been successful in reducing the noise of paper-making machines and was doing some consulting work for Caterpillar. In August 1949 we signed a contract with Silencing Consultants of America for a special noise analysis of our tractors.

We sent a model 77 tractor to Toledo for the evaluations. Mori, an engineer with Silencing Consultants, had a special treadmill made so he could operate the tractor in each gear. He became confused when he changed gears and got as many as eight new frequencies. He ignored our suggestion to consider the frequencies being the number of teeth in mesh per second as the predominate source of sound generated by the transmission. His data were within the variability of his equipment and his conclusions were false. We were running out of funds in the contract during early March 1950 so we gave notice that the project in Toledo would be canceled as soon as certain data were obtained.

We brought the 77 tractor to Charles City and purchased a high-quality tape recorder, sound analyzer, frequency oscillator, and other equipment to do our own sound analysis. The results of these tests revealed that spur gear noise could be reduced by more accurate machining and fitting the gears. It was amazing how much the sound was reduced after the transmissions were worn in. We started crown-shaving the gears in about 1950, which assisted in a quick wear-in of the transmissions.

Mufflers

Oliver was the first to use aluminized steel mufflers. We had gotten some complaints from dealers that the mufflers were wearing out too soon and this was a warranty item, so we looked for a longer-lasting material. We consulted steel companies who told us that they were experimenting with aluminized steel. Nelson Muffler Co. of Stoughton, Wisconsin, purchased the aluminized steel and made

When Oliver put aluminized steel mufflers on the Fleetline tractors, it was the first time they were ever used. They were introduced to provide a longer-lasting muffler because mufflers were a warranty item that had to be replaced by dealers when they wore out too soon. The small aluminized mufflers on the Fleetline tractors were placed under the hoods.
T. Herbert Morrell collection

mufflers for Oliver. Later, when the automobile industry started using aluminized steel mufflers, the supply of aluminized steel was greatly reduced. We were asked to accept a substitute. We argued, but did not win. We substituted regular steel painted with high temperature silicone paint until aluminized steel was available in larger quantities. Donaldson also made mufflers and was the second source.

Axles and Hubs

At the start of the Fleetline tractor production, Oliver used a very high-grade alloy steel (A - 4140) for rear axles. After machining, the axles were completely hardened to approximately 340 Brinell hardness. This material and heat treatment was generally the selection for axles in the tractor industry and had been used quite successfully in the older 60, 70, and 80 models.

During the late 1940s, the harvesting products plant at Battle Creek, Michigan, developed a two-row mounted corn picker to mount on the model 77 and 88 tractors. These pickers were excellent and they created much enthusiasm among Oliver dealers. The next year it was full speed ahead for the manufacture of the Oliver No. 4 mounted corn picker.

A pilot run of approximately 50 No. 4 mounted corn pickers was sold to owners of 77 and 88 tractors within the Corn Belt. Customers with large fields of corn were chosen wherever possible to get as much usage and field experience as possible. Also, the farms where the pilot units were placed were carefully chosen so that a representative from Oliver could follow each unit.

Some rear axle failures occurred on the 77 when the No. 4 picker was installed. Oliver's evaluation revealed that the rear axle failures on the 77 with the No. 4 picker did not occur in the fields, but on roads, especially on gravel roads with a rough, washboard-like surface. The early failures were on 77 tractors owned by custom operators who picked corn for neighbors and then drove over the rough roads at maximum speed to the next field. The rough roads caused some high shock loads on the rear axle.

A failed 77 rear axle was sent for Oliver's Charles City Engineering to evaluate. We noted that there were scratches on the

Early failure of the rear hub on prototype Model 88s caused Oliver to switch to a malleable cast-iron axle and to use an axle hub with three U-bolts. There were no more failures of these hubs. To provide better traction, Oliver used steel rims attached with bolted lugs to cast-iron wheels. ***Jeff Hackett***

curved part between the anti-friction bearing surface and the larger part of the axle. The failure was typical of bending fatigue.

The Charles City engineering department with the assistance of Oliver's metallurgists and factory industrial engineers brainstormed about the situation. First, we needed a severe test setup whereby we could obtain a failure quickly on a 77 production rear axle. We would then use this same setup to test any proposed solution. A test fixture was built which was capable of adding weight equivalent to the tractor and the corn picker with additional weight as a severity factor. We decided on a weight severity factor of one and one-half that of the tractor and the picker. We drove the axle at the speed of an electric motor as another severity factor. If an axle on a 77 tractor was driven at this high speed, it would be travelling about 140 miles per hour.

Second, we needed to provide a smooth radius which located the outer axle bearing to a shoulder backed by the larger part of the axle. Scratches on the surface of a radius of this type will tend to cause early fatigue failure. While we were testing a production axle to establish a base for comparison, we rolled the radius which located the outer axle bearing. The purpose was to remove any machine scratches on the test axle.

The first failure was on the tapered outer axle bearing. The severity factors had exceeded the deflection permissible for such a bearing, so we substituted a barrel roller-type outer axle bearing that would tolerate more deflection or misalignment. Production axles then failed consistently after about four hours of test operation.

The axle with the smooth radius, which located the outer bearing, showed some improvement over the production axle, but this was not an adequate solution. Our goal was to find an axle that would pass our severe test without failure after four million revolutions. Metallurgists had determined that if a part had undergone four million bending cycles without failure, the part would not fail by fatigue unless the load on the part was increased.

The Oliver metallurgists at Charles City were working with induction-hardening equipment for large parts such as axles. Induction hardening is a process whereby high-frequency electricity is applied to heat-selected areas of a part. When the part is cooled quickly, the hot areas become very hard. We asked for engineering representatives from steel companies to meet with us. We selected a medium carbon steel (C-1038) and then induction-hardened the outside diameter of the axle to approximately 3/16 inches deep. The outer and inner ends were then ground to size to fit the outer and inner bearings.

These axles did not fail by fatigue under our severe test conditions. Only a few days were required for the tests and to start making replacement axles from C-1038 induction-hardened steel. We had great cooperation from our metallurgists, the steel companies, bearing companies, our industrial engineering, and our machine shop.

Our service department provided the new axles free to the customer having a 77 tractor and No. 4 picker. We also paid a labor allowance for the time to exchange the axles. A program was established to change any inventory 77 tractors sold with a No. 4 picker to be attached.

Changing from the A-4140 steel completely hardened to C-1038 induction-hardened axles that had only the outer surface hardened, was a significant cost saving amounting to many dollars per tractor. The solution of this problem is an excellent example of problems becoming the mother of invention. When there is no problem, we tend to continue in the same way.

The rear axle hub on the experimental tractor had a wedge-shaped projection which fit tightly into a wedge-shaped groove in the axle. Two U-bolts wrapped around the axle and then passed through some holes in the hub. The U-bolts were to draw in and lock the projection on the hub into the groove of the axle. The experimental hubs were very carefully machined by hand. There were no hub failures on the experimental tractors, but in late 1945 and early 1946 there were a few hub failures on the pilot run of 88s that occurred a few weeks before the start of the first 88 production run. We decided to make the hubs of malleable casting instead of the regular high-strength cast iron. Some of these tractors had malleable cast rear axle hubs. These could be identified by the thinner flange and other areas of the casting.

Oliver engineering searched for and found a rotating shock load test that would cause a cast-iron hub to fail quickly. A movie camera

The steel on the rear axles of the Fleetline tractors was hardened on the surface by heating with electric current, called Induction hardening. This provided a more satisfactory axle and solved a number of problems in which rear axles had failed under heavy loads. ***T. Herbert Morrell collection***

was used to film the end of the axle and the hub during the test. The film showed that just before the failure occurred, the hub separated from the axle in the vicinity of the wedge-locking key between the axle and the hub. We then tested a malleable cast-iron hub. It was an improvement over the regular casting but not enough to assure us that it was the answer to hub breakage.

We brainstormed about the situation and all engineers involved agreed that if we clamped the hub and axle together tightly enough that there would be no separation during these high-shock loads, the cast-iron hub would not fail. We decided to try three U-bolts instead of two. The addition of the extra U-bolt was tested and we had no failures during the severe shock loads. Incidentally, these tests were conducted without an operator in the seat but with a remote tractor ignition shut-off because the safety of the operator was most important. To our knowledge, there were very few, if any, hub failures with the three U-bolts tightened to engineering's specifications.

Steel wheels were purchased from Electric Wheel Co. of Quincy, Illinois. For lighter-weight tractors that did not need any extra weight for traction, the steel wheels were manufactured in two parts—a disc welded to a rim. The tread adjustment was limited but quite satisfactory for most applications of Standard tractors. For higher traction requirements, such as for Row Crop tractors, Oliver manufactured cast iron-wheels that provided additional weight for traction. These Oliver cast-iron wheels were machined to provide many adjustments for tread width. The rim was separate from the wheel with formed ridges that were offset from the center line on the inside diameter of the rim. The separate offset rim with the cast-iron wheel and straight adjustable axle provided for great tread adjustment. The steel rim was attached to the wheel by positioning the rim on the cast-iron wheel and then using bolted lugs to hold the rim ridge tightly on the wheel.

Tires

Tires were provided by Goodyear, Goodrich, and Firestone. The cost of the tires was a major factor in the overall cost of a tractor. Goodrich had a unique policy in which the company assigned sales representatives to individual companies. Stan Murray had only one account and was their sales representative to all of Oliver plants and sales branches. In contrast, there were many sales representatives from Goodyear and Firestone during the development of the Fleetline tractors and successive years.

Brakes

The Oliver model 60 tractor had an equalizer brake pedal system designed by Milford D. Stewart. He received patent number 2,443,331, which was assigned to Oliver. The left and right foot pedals were located above the transmission and could be operated by the right or left foot. A bar was attached to both brake pedals with a pad in the center. The operator could assist tractor steering by putting pressure on the brake pedals. Pressure on the right pedal would assist in right turns and pressure on the left pedal would assist in making left turns. Pressure on the pad of the equalizer bar would set both brakes to stop straight ahead movement of the tractor. This brake control system was so successful that it was used on the new Fleetline models.

Equalizer brake pedals were offered on the Fleetline series. This allowed the operator to brake only the right or left rear wheel to aid in turning the tractor. The band brakes used on the early Fleetline tractors were less effective when the tractor was going in reverse and had some experimental failures. This led Oliver to contact brake companies and become the first to adapt the new double-disc brakes to the Fleetlines. They were introduced in late 1950 and early 1951.
Jeff Hackett

The experimental, pilot runs, and first production runs of the Fleetline 66, 77, and 88 tractors had a band brake that had been used successfully on the Oliver model 60 tractors. It was self-energized in the forward direction. That means when either the right or left brake was engaged in the forward direction, the band would wrap tighter around the drum and add to the braking capacity. The disadvantage was that this type of brake was less effective in reverse. We started getting complaints about the lack of braking capacity in reverse.

To remedy this, we tried anchoring the brake band by a lug on the band that fit into a slot on the rear main frame and brake cover. This provided some self-energizing effect in both forward and rearward directions. About this time, I was traveling with a young New Jersey territory manager by the name of Dutch Zandbergen. We stopped to see a large tomato-farmer who had brake problems and found him in the middle of a very large tomato field. Dutch introduced me and said to the farmer, "You said that you would like to get your hands on an engineer from the factory and here he is." He jumped off the tractor, started swearing at me, lost his hat in the wind and yelled at his son to retrieve his hat.

The tomato farmer used his brakes frequently and failure of the band and friction material were causing problems for him. We took brake bands off one of his dealer's new tractors to help him until a better brake design could be provided.

Shortly after the New Jersey trip, engineering was asked to send someone to see our dealer Johnny Wheeler in Lubbock, Texas. His customers had experienced some brake failures while operating on the level plateaus. We examined the failed brake parts and then went to see some customers. The farmers had learned that if the tillage was contoured rather than done in a straight line, the resulting eddy currents would reduce or nearly prevent wind erosion. Because the very wide implements being used tended to force the tractor to go straight ahead and the shear resistance of the soil was not sufficient to steer with only the front wheels, it was necessary use the brakes to steer the tractors on the contour surface. The nearly constant braking required in these circumstances caused the friction material and bands to fail frequently.

We brainstormed about this problem and agreed that we should set up a test that would duplicate the field failures. We established a figure-eight test area on some concrete. The steering wheel was tied so that the steering had to be done only by applying the brakes. The operator traveled 50 feet, steered to right angle turn, traveled 50 feet, steered to the left, continuously in a figure-eight pattern. After approximately 30 minutes of the test, the binding material in the brake lining was failing. The brakes got very hot and it did not take too long to duplicate what was happening in the field.

We decided that we must provide a brake that would dissipate the heat as fast as the heat was generated and where the temperature created under these figure-eight test conditions would not exceed the failure point of the brake lining. We invited brake companies to observe our tests and then provide us with some brakes to test. We tried some proposed brakes and all of them showed some improvement over our production brakes but none met our requirements.

We finally contacted Dent Parrott of Auto Specialties Co. in St. Joseph, Michigan. He became very interested in our problem. We worked with him on the application of a double-disc brake to our 88 tractor. That brake showed very promising results during the figure-eight test. We continued to test, evaluate, and improve until we arrived at the brake that met our requirements. These tests occurred in 1949 and early 1950. We started producing Fleetline tractors with these double-disc brakes in late 1950 and early 1951.

Electric Lights

My family bought a regular Farmall tractor in 1927. We farmed more than 360 acres with this small tractor, horses and mules. In 1928, my brother Lloyd and I equipped the tractor with electric lights so that we could operate the tractor nearly 24 hours each day. Using clamps, we mounted a generator from a 1924 Chevrolet 490 on the right side. The generator was driven by the belt pulley shaft. We used another set of clamps to mount the generator on the drawbar and drive it from the power take-off shaft when the tractor was used for belt work. Our 6-volt automobile radio battery was charged by the tractor generator off the lighting setup. We bolted angle irons to the left and right drive housings located inside each rear wheel. We used four Dodge headlights, two mounted to the angle irons facing forward, one on the right side facing rearward,

Lights were mounted on the wheel guards, as shown on this Row Crop 88. Lights failed on the experimental Fleetline tractors because the wires were torn loose in the field. To prevent this from occurring, the wires were run through steel tubes attached to the wheel guard and the axle housing. ***Jeff Hackett***

and the fourth headlight on the Farmall steering sector which was located in the front, ahead and above the tractor radiator. This fourth headlight turned when the tractor was steered and lighted the area where the tractor was turning.

The early Farmall tractors had spur-gear sector-pinion steering which created some dangerous problems. The front wheels were steel with an angle iron offset from the center line of the vertical steering shaft. If the front wheel struck the edge of a stone below the surface of the ground, the steering wheel would start rotating at high speed causing some skin to be removed from the operator's knuckles. Also, the operator had to react quickly to prevent the tractor from turning on top of the implement.

I wrote a short letter to IHC saying, "I am just a farm boy and have had some steering problems on your Farmall. Why don't you use a worm pinion and sector that will not reverse and cause the operator to lose control of steering?" I never received a reply from IHC, but the IHC F-12, F-20 and F-30 tractors introduced in the early 1930s had the worm pinion and sector enclosed with its own lubrication. We traded for an F-30 for more power. We transferred the lighting from the original Farmall except the fourth light which could not be used on the F-30 because of the enclosed worm pinion and sector.

Oliver first offered electric lights as an option on the Oliver 70 Row Crop tractors in 1935. The front lights were located on each side of the grill. A rear light to illuminate implements was located on the right rear wheel guard and facing rearward. The Oliver 60 had forward and rearward working lamps located on the right wheel guard. One forward working lamp was mounted on the left wheel guard.

The Fleetline 66, 77, and 88 followed the Model 60 light locations. The advantage of these locations was that they provided more flood lighting on front-mounted working implements, such as cultivators.

One of the first problems of the test tractors during late 1944 and early 1945 near Phoenix, Arizona, was lighting failure after a short time of operation. The long staple cotton stalks tore the light wires from the tractor. The wires were installed between the main frame and the wheel guard, and the lights were located on some temporary wheel guards that were designed for some older tractor models.

To solve this problem, I designed a wiring setup whereby the wires from the light switch were mounted on the inside of the operator platform flange to the rear axle area. Then the wires passed through a steel tube attached to the rear side of the axle housing and the inside of the wheel guard. The light wires were then attached in a protected area of the wheel guard and extended to the lights mounted on the wheel guard. This improvised wiring on the experimental tractors was successful so nighttime operation in the Phoenix area could continue satisfactorily.

For production models, grooves for the light wires and a recess for the cover plate were added to the rear side of the axle carrier's hexagon shape. Both the tractor and implement engineers were involved in the selection of the hexagon shape of the axle carrier. This shape became a very accurate and solid mount for implements. The production light wires were mounted on the inside of the operator platform flange. A steel tube protected the wire to the recess in the axle carrier, the wire passed through the recess, through the axle carrier flange and then inside the wheel guard support channel to the light and light bracket.

The production light bracket was designed as a steel stamping with strengthening ribs. It was attached to the wheel guard by bolts through the wheel guard and through the support channel. The result was a very sturdy support for the light. Fatigue failure of the wheel guard on some competing tractors was experienced because the lamp bracket was attached to the wheel guard with insufficient support. The thin metal around the lamp and its bracket would crack. Such a failure did not occur on the Oliver Fleetline tractors.

Ridemaster Seat

A main consideration in the development of the famous Fleetline 66, 77, and 88 tractors was to provide an easy riding seat. Oliver conducted much research on the riding capabilities available for seats in trucks, automobiles, and other vehicles. Bostrom Manufacturing Co. in Milwaukee, Wisconsin, was a leader in the research for easier riding seats for trucks and off-road vehicles. This company had

A common failure on earlier models was the cracking of the wheel guard around the lamp bracket. The Fleetline tractors were all released with an improved wheel guard and cast-iron lamp bracket that avoided this problem. ***Jeff Hackett***

developed an in-depth research program to improve riding comfort. They became quite interested in Oliver's seat project and conducted studies on farm equipment operating in the field.

Many designs were considered and some were tested in cooperation with Bostrom. One used a modified shock absorber but it did not provide the desired smooth ride. The design chosen was a seat mounted on torsional rubber springs. Much data was obtained on the movement of the vehicle versus the movement of the farmer. With this unit, there was very little up and down movement of the farmer when the tractor ran over very rough ground. Tests continued on Oliver's experimental tractors and on farmers' tractors. Oliver's Ridemaster seat was introduced in 1948 with the introduction of the Fleetline tractors.

The new seat suspension system was made by Bostrom, the supports were manufactured by Oliver, and the steel stamped pan and the cushion were purchased from other sources.

It was 1960 before John Deere introduced tractors with an easy riding seat. Other companies started providing an easy riding seat on their tractors, but all were much later than Oliver.

OLIVER STANDARD "88" SPECIFICATIONS

(Subject to Change Without Notice)

ENGINE: Six-cylinder, four-cycle, vertical, valve-in-head. Engine speed 1600 r.p.m. Bore, 3½"; Stroke, 4". Displacement, 230.9 cubic inches. Four main bearings, removable precision-type, steel back, babbitt lined, with shims for easy adjustment. Connecting rod bearings, removable precision-type, steel back, babbitt lined, with shims for easy adjustment. Pistons, nickel iron, with three compression and one oil ring above the pin. Removable wet-type nickel iron cylinder sleeves. Heavy crankshaft dynamically balanced. Crankcase ventilation by breather on top of rocker arm cover. Water temperature and oil pressure gauges. Electric starter regular equipment.

IGNITION: Modern battery-type distributor, sealed against dust. Centrifugal automatic spark control at all engine speeds.

GOVERNOR: Centrifugal, variable speed type. Fully enclosed and automatically lubricated. Hand control at steering wheel.

LUBRICATION: Engine lubrication is by means of pressure from large capacity oil pump located in sump. "Floato" screen oil pump inlet. Oil capacity in crankcase, 6 quarts. Oil capacity in transmission and final drive, 4 gallons. Chassis lubrication by grease gun through pressure fittings.

OIL FILTER: Improved type. Element easily replaceable and filter base attached directly to crankcase. No tubes or fittings.

AIR CLEANER: Oil wash type. Cup easily removed for cleaning.

COOLING SYSTEM: Full length water jacket for uniform cooling. Water circulation by pump on fanshaft. Water temperature controlled by a by-pass type thermostat. Fan and pump directly driven by V-belt from crankshaft. Fan belt is tightened by increasing effective diameter of driving pulley which results in improved cooling as belt wears and requires adjustment.

FUEL SYSTEM: Gravity from fuel tank. Capacity of fuel tank, 20 gallons. Fuel strainer and removable sediment bowl between fuel tank and carburetor.

CLUTCH: Single plate, dry-type, 10 inches in diameter. Self-adjusting spring loaded and foot pedal operated.

TRANSMISSION: Selective sliding spur gears, alloy steel, carburized and hardened. Shafts are heat-treated alloy steel, mounted on ball and roller bearings. Six speeds forward; two reverse. Transmission fully sealed and running in oil. Special combination for applications requiring it—four speeds forward, four reverse.

SEAT: New Ridemaster adjustable seat. Pan-type. Rubber torsion springs. Cushion shown, slightly additional.

NEW . . . RIDEMASTER SEAT!

Now, Oliver offers as standard equipment the most comfortable seat ever placed on a farm tractor. Every jolt is cushioned by large rubber springs to give you a "floating" ride that makes driving a pleasure, even over the roughest ground. The RIDEMASTER enables you to stay on the job longer, do more work per hour, and end the day with less fatigue.

This new RIDEMASTER seat is fully adjustable to furnish equal riding qualities for a slight farm youngster of 100 pounds to a 275-pound husky. It can also be adjusted fore and aft to suit leg length. Cushion shown, slightly additional.

BRAKES: Differential, external contracting type. Fully enclosed, external adjustment. Foot operated, either individually or equalized.

POWER TAKE-OFF: (Special Equipment) Direct-drive-type, independent of main clutch. Shaft 1⅜" diameter, 6B spline. Rotates clockwise at 533 r.p.m. Located on centerline of tractor in separate housing with own clutch. Safety shields.

STEERING: Worm and gear type. Enclosed and operated in oil.

WHEELS & TIRES: Front wheels two cast semi-steel, demountable rims, 6.00-16 tires, mounted on tapered roller bearings. Rear wheels two cast semi-steel, demountable rims, 12-26, 13-26 or 14-26 tires. Front wheel tread 48¾" to 54¾". Rear wheel tread 54" or 62". (See Instruction Manual for tread adjustments.)

SPEEDS—Forward: First, 2⅓ m.p.h.; Second, 3¼ m.p.h.; Third, 4⅛ m.p.h.; Fourth, 5⅓ m.p.h.; Fifth, 6¾ m.p.h.; Sixth, 11¾ m.p.h.; Reverse low, 2½ m.p.h., Reverse high, 4⅝ m.p.h. These are rated speeds of tractor at rated 1600 r.p.m. engine speed and 13-26 tires.

DRAWBAR: Adjustable. Vertical adjustment 7¾" to 18⅛"; five fixed positions. Lateral adjustment, 20⅝" total.

BELT PULLEY: 11⅜" diameter, 7¼" face, 992 r.p.m., 3080 ft. per minute. Pulley or entire mechanism removable.

GENERAL DIMENSIONS: Length overall, 133⅓ inches. Width overall, 68 inches. Height to top of radiator, 57¾ inches. Wheelbase, 79⅝ inches. Ground clearance, 13¾ inches. Approximate field weight, 4950 lbs. (Weight does not include wheel weights, liquid ballast or extra equipment.) Weights and dimensions given are for 13-26 tires.

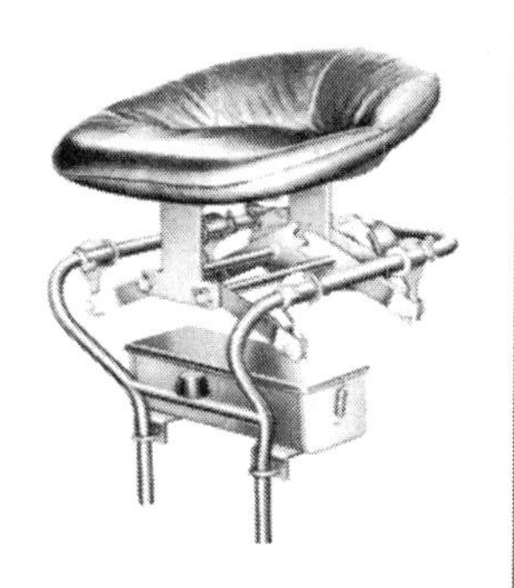

This specification sheet for the Oliver Standard 88 features the new Ridemaster Seat, a feature that was part of increasing awareness of the importance of operator comfort. *Floyd County Historical Society*

The Fleetline series of tractors had a number of features that reflected Oliver's commitment to operator comfort. One of the areas in great need of improvement was the seat. This photograph shows the old-style rigid, canvas-style seat, which was very uncomfortable for the operator. *T. Herbert Morrell collection*

One of the design criteria for the Fleetline tractors was to provide more comfort for the operator. This Ridemaster seat with rubber torsion support was designed to do just that and was very popular with the customers. ***T. Herbert Morrell collection***

Carburetors

Carburetors were purchased from Carter and Marvel Scheibler. Air cleaners were purchased from Donaldson in the Twin Cities area at the start of the Fleetline tractor development. Then they expanded their manufacturing facilities to Iowa. Bob Larson was the sales representative who worked very closely with tractor manufacturers. There was never a dull moment outside of business when Bob was present. One of his favorite expressions was, "Call me S.O.B., Sweet Old Bob." Some of the first dry-type air cleaners were purchased from United Air Cleaner.

Cooling Systems

Radiators were purchased from Young, Modine, and other companies.

Quality Hardware

Hardware was very important for the overall quality of a tractor. At one time Oliver purchased both non-hardened and hardened capscrews and bolts of the same size. A check on the assembly line indicated that some soft capscrews were being used where hardened capscrews were specified and vice versa. This problem was solved when the Charles City plant decided to use only hardened hardware. This eliminated extra purchasing, storage, handling, and other hidden costs of having extra parts in inventory, so it became economical to purchase only the hardened hardware.

Bearings were purchased from various sources. Timken Bearing Co. of Canton, Ohio, and Bower Roller Bearing Co. were the sources for tapered roller bearings. Tapered roller bearings were used throughout the transmission, differential, and final drive. We had a long-term relationship with Timken.

Hyatt Bearing Co. had a competitive tapered roller bearing except that the rollers were barrel shaped. The advantage claimed by Hyatt was that their bearings would accept more deflection of the shaft or axle on which the bearing was mounted.

Ball bearings were purchased from New Departure and Bearing Co. of America. The ball bearings had advantages for installation on some shafts in optional drives such as the power take-off.

Straight roller bearings were purchased from Torrington Bearing Co. and McGill. When I met Kip Recor, he was on his first trip as an engineering representative of Torrington and I had just completed the power take-off design in 1945. I needed a needle bearing for the

This 1950 Standard 88 demonstrates the long-lasting nature of Oliver tractors. After more than 46 years, it has not been restored, but is still being used. ***T. Herbert Morrell collection***

One of the companies Oliver descended from, the Nichols and Shepard Threshing Machine Co. was founded in 1848, so 1948 marked 100 years of manufacturing for Oliver. The introduction of the Fleetline 66, 77, and 88 became part of Oliver's Centennial Celebration. These new tractors were displayed at many state, county and local fairs. Trade magazines used color pictures and listed the Oliver innovations in their advertising. Color brochures and powers, such as the "Three Beauties" shown here, were available to Oliver's dealers for distribution to potential customers. ***T. Herbert Morrell collection***

output shaft to provide clearance for the clutch. Torrington was able to supply it, so Kip got his first bearing application. He was very happy and we became good friends.

Switches, steering wheels, wiring harnesses, gauges, lights, and a large number of items were also purchased.

Versatility

The versatility of the 66, 77, and 88, when designed together, created a custom-built combination. Many special assemblies of each model could be provided without designing many, if any, new parts. The Oliver Fleetline tractors brought a host of improvements to the farmer, many of which were incorporated by competing manufacturers.

Introduction of the Fleetline: Oliver's Centennial Celebration

Two of Oliver's original companies were founded in 1848, the Nichols and Shepard Threshing Machine Co. and the American Seeding Machine Co., so 1948 marked 100 years of manufacturing by Oliver. The introduction of the Fleetline 66, 77, and 88 became part of Oliver's Centennial Celebration. These new tractors were displayed at many state, county and local fairs. Trade magazines used color pictures and listed their innovations in their advertising. Color brochures and posters, such as the "Three Beauties," were available to Oliver's dealers for distribution to potential customers.

CHAPTER THREE

XO-121 Research Program

The XO-121 was a very important research program for Oliver. The engine used a raised compression ratio and specially developed fuel to dramatically improve the power output and efficiency of comparable tractor engines. The research performed on this engine was used in the development of later Oliver production engines.

By 1953, the Ethyl Corp. became quite concerned about the success of the Oliver diesel tractors and the future effect on the use and sale of tetraethyl lead for anti-knock purposes in gasoline. If all agricultural tractors, automobiles and other similar applications changed to diesel, the amount of tetra-ethyl lead sold would be greatly reduced.

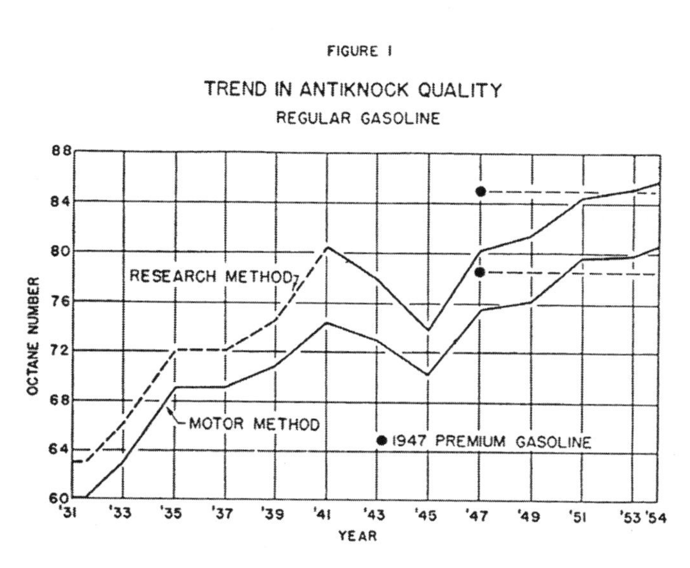

Figure 1. Trend in Anti-knock Quality: Regular Gasoline. ***T. Herbert Morrell collection***

The research performed with the XO-121 was applied to later Oliver engines, especially the engine for the 1800 and 1900, which used a raised compression ratio and concave piston crown to increase performance and efficiency. The Oliver Super series tractors also benefited from the XO-121 program research. ***Jeff Hackett***

Oliver was looking for improved tractor performance, especially lower fuel consumption with higher grade gasoline. So, Oliver and Ethyl developed a joint research program called the XO-121, X for experimental, O for Oliver, and 121 for a 12:1 compression ratio. But there was a problem—the board of directors of Ethyl Corp. did not want to approve the experimental program.

Charles F. Kettering of General Motors had directed the development of a high compression automobile research engine that showed much promise in making gasoline engines more efficient. He was interested in the XO-121 project because he wanted to make gasoline engines more fuel efficient, comparable to the tractor diesel engines. Kettering finally convinced the Ethyl board of directors to authorize the XO-121 project after Ethyl's laboratory management had failed to sell the project to the board.

The 16mm film, *Getting Ahead of Tomorrow*, was about the development of the XO-121. This is a scene in the film where executives of Ethyl and Oliver discuss plans for the research tractor with a compression ratio of 12:1, which became the XO-121. Pictured, from left, are Rollin Gish, Richard Scales, George Bird, A. King McCord, Bynum Turner, Herb Morrell, and J. B. Macauley. Gish, Scales, Turner, and Maccauley were with Ethyl Corp. and the others were with Oliver.
T. Herbert Morrell collection

Executives of Ethyl and Oliver cooperated in the research program. Ethyl representatives provided the experimental gasoline and monitored the design program at Charles City. Ethyl also agreed to conduct some engine tests in their laboratory in Detroit, Michigan. Oliver's Charles City engineering department was responsible for the engine and tractor design and special parts. In fact, the project became so important to the two companies that a film was made about it, *Getting Ahead of Tomorrow*.

An initial challenge was determining how much strength would be required in the crankcase assembly. We chose a four-cylinder Hercules diesel engine with multiple main bearings as a base for the engine design. We then proceeded with the design of the cylinder

The right-side view of the XO-121 engine shows most of the accessories, such as the alternator, distributor, and the ignition coil. The crankcase was taken from a four-cylinder Hercules diesel engine, with the top end of the engine constructed by Oliver. The diesel crankcase and the XO-121 cylinder head made a good combination. *T. Herbert Morrell collection*

head, manifolds, carburetor and other related parts. A special tractor front frame design was designed to accommodate the Hercules basic engine mounts and align with the tractor transmission input shaft. The radiator, air cleaner, sheet metal, and others were also Oliver's design responsibility, as well as the assembly and field tests of the final tractor. Charles Van Overbeke and Walter Roeming were the Oliver engineers responsible for the engine design. H. (Heine) Mueller and Keith L. (Punch) Pfundstein were the Ethyl Corp. engineering representatives. Rollin Gish was in charge of the engine test program at the Ethyl Corp. laboratory in Detroit, Michigan.

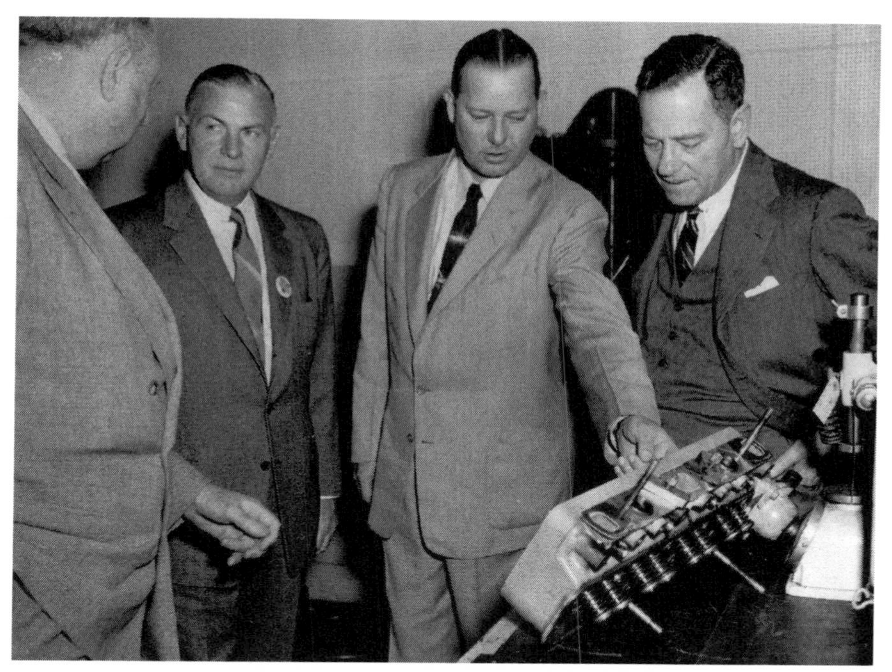

(from left) George Bird, A. King McCord, and Herb Morrell of Oliver, and J. B. Macauley of Ethyl Corp. review details of the XO-121 cylinder head. The shape and contour of the combustion chamber were crucial in obtaining the desired results. *T. Herbert Morrell collection*

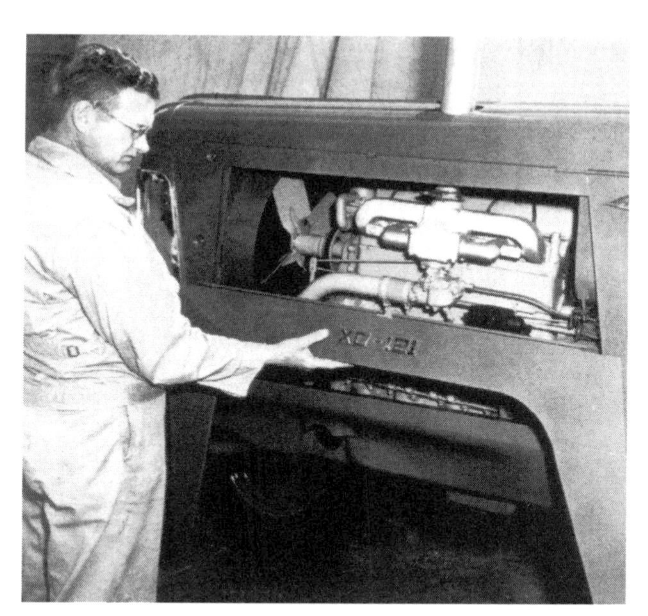

Charles City engineering was responsible for producing the experimental tractor. Max Denham of the Experimental Engineering Department is shown here with the XO-121. *T. Herbert Morrell collection*

The XO-121 was tested by plowing a field with Max Denham of Oliver Engineering as the driver. This and other tests in Charles City proved that the 12:1 compression ratio was practical if gasoline of the required high quality became available. *Floyd County Historical Society*

During May 1954 the XO-121 tractor was shipped to the General Motors Proving Grounds near Milford, Michigan, to be demonstrated for Ethyl and Oliver executives. Keith Pfundstein met Oliver's executives, board Chairman Alva Phelps and President A. King McCord, at the Detroit airport and drove them to the GM Proving Grounds. Phelps argued that the XO-121 was a waste of Oliver's resources and a waste of his time because it had no purpose and so proved nothing. McCord tried his best to convince Phelps to be patient and open-minded. Both Phelps and Kettering had come from General Motors and McCord reminded Phelps that the whole world knew what Kettering's high-compression Oldsmobile engine at 12:1 compression ratio had meant to the industry.

When Phelps and McCord arrived at the proving grounds, Heine Mueller and Rollin Gish were plowing with the XO-121 tractor. The photo shows A. King McCord, driving the tractor with Rollin Gish riding with them to answer any questions that either of them may have had. When the tractor pulled up in front of the group, Phelps walked to it, gingerly touched the muffler and said, "It can't be very hot because at 12:1, there can't be much heat left over."

The tractor performed exceptionally well. At the conclusion of the demonstration, Phelps was in favor of continuing the XO-121 program. After the demonstration, Pfundstein drove Phelps to the airport. Phelps was quite talkative and friendly, and apparently was pleased with what he observed at the demonstration. They were discussing research and application of new technology when Phelps uttered one of his famous expressions, "It's amazing what foolish thoughts I get when I think too long alone."

The program developed as planned and with results beyond our expectations. There was 43 percent more power on 28 percent less fuel, and the brake-specific fuel consumption of 0.385 pounds per horsepower hour was equal to that of the best diesel engines available.

During the 1954 American Society of Agricultural Engineers meeting at the University of Minnesota, I presented a technical paper titled "Looking Ahead of Tomorrow in Tractor Engine Design" with Homer Dommel as co-author. There were many

A. King McCord is driving the XO-121 at the General Motors Proving Grounds with Rollin Gish riding along to answer any questions about the tractor and tests. ***K. L. Pfundstein collection***

questions and good discussion. My friend Stanley Madill, a top engineering executive of Deere and Co. made an interesting comment to me afterward. He said, "You will never know what you have just done for John Deere Engineering." His meaning became evident when John Deere introduced its new line of high-compression tractor models 1010, 2010, 3010, and 4010 in 1960 with vertical multiple-cylinder engines making their old two-cylinder engines obsolete.

Oliver was host to a luncheon after the paper was presented. The guests included the media from newspapers, radio and TV; engineers and executives of the U. S. tractor companies; and ASAE staff members. The luncheon speaker was Charles F. Kettering of General Motors, considered the Thomas Edison of the automotive industry. He was the owner of approximately 200 patents of which the electric starter is a good example, owned a large amount of GM stock, and was a major contributor to the Sloan-Kettering Cancer Research Center in New York City, New York. He was in demand as a public speaker because he always held his audience spellbound.

Kettering told a story during the luncheon about his development of 10- or 12-inch diameter pistons for the Electromotive Division of GM. He called some engineers for a meeting and asked, "Why hasn't anyone made aluminum pistons for high-speed diesel engines?"

The answer was that aluminum was a softer metal and would not withstand the high pressures of diesel combustion. He said that he was tired of his staff of young bright engineers telling him that everyone knows that an aluminum piston that size can't survive the high pressures and temperatures of a diesel engine. He told them that he did not want to hear that again unless you yourself have been a piston in a diesel engine. The aluminum would just need some reinforcing ribs. He said, "The aluminum didn't know that it shouldn't work so it soon became the preferred material for high-speed diesel engine pistons."

He related that he had a meeting with representatives from the American Petroleum Institute regarding the future for improved gasoline quality. He was told the high-quality gasoline cost would be prohibitive. He then asked, "If that is the case, maybe we are headed in the wrong direction. How about making 60 octane gasoline?"

He was told that the cost of producing 60 octane gasoline would also be prohibitive. Kettering replied, "Then what in the hell is the relationship between octane rating and the cost of gasoline?"

Kettering was very interested in the XO-121 program and was kept informed of its progress. During the luncheon meeting, he said, "I have made some evaluations and if all gasoline engines had this high compression ratio and the gasoline quality to match, we could stack railroad tank cars full of gasoline from Chicago to New York and that would be our yearly saving of gasoline."

Research on gasoline engine efficiency continued based upon what was learned from the XO-121. This information became very valuable to Oliver Engineering at Charles City as a consideration for future tractor engine design.

In September, 1961, I presented the Society of Automotive Engineers (SAE) a technical paper "Development of Oliver's New Gasoline Engine," co-authored by K. S. Minard, during the Heavy Duty Vehicle Meeting in Milwaukee, Wisconsin (Appendix C). Super 88 gasoline engines were modified and tested in many ways over several years. There were many things to consider as described in the paper. Each item of change was studied and tested individually. Some of these tests required and consumed several hundred dynamometer hours.

Figure 13 of Appendix C shows the final combustion chamber configuration which is a combination of the XO-121 and a concave piston head. This combustion chamber was introduced on the 1800 tractor in 1960. Nebraska test number 766 on the model 1800 established a new all-time low fuel consumption record of 0.472 pounds per horsepower hour, or 13.18 horsepower per hour per gallon on the maximum-power of the power take-off test. As far as we know, this record has not been broken at any official test laboratory. The paper indicates the in-depth study needed to design a tractor engine that would most effectively utilize the quality of gasoline available when the 1800 tractor was to be introduced in 1960.

The XO-121 tractor was given to Iowa State University for display. The university kept it for many years and then gave it to the

Living History Farms west of Des Moines, Iowa. The tractor was repainted and almost lost its identity. The first president of the Hart-Parr Oliver Collectors Association, Wayne Wiltse, was successful in getting the XO-121 tractor from the Living History Farms and brought to Charles City where it was restored to its original colors. It is now a part of the Floyd County Museum in Charles City, Iowa.

CHAPTER FOUR

The Super Tractors

Super 55

During the 1940s, Ford Motor Co. was selling a large number of model 8N tractors. Alva Phelps, Oliver's CEO, established a mandate for Oliver to copy the 8N tractor. The Ford-Ferguson 9N preceded the 8N and was introduced in 1939, but around 1947 Ford and Ferguson dissolved their partnership and Ford continued with basically the same tractor called the model 8N. Ferguson sued Ford for patent infringement and sought damages of about $340 million. During the early 1950s, this lawsuit was settled with Ford paying about $9 million to Ferguson. Those of us in Oliver's Charles

This XO-67 tractor was the prototype for the Super 55. Six were produced in June 1953. ***Carl Rabe***

City engineering department expected Ford to release a new, more modern tractor that would avoid infringing upon Ferguson patents. We advised Oliver not to copy an old, obsolete tractor, but Oliver board chairman Phelps would not change his mandate because he thought Oliver could sell such a tractor in large quantities.

To avoid the Ferguson patents, Oliver Engineering worked with Vickers to provide a draft control valve to be compatible with its continuous running hydraulic pump. The rest of the tractor was basic and very similar to the Ford 8N. We built a prototype to review with Alva Phelps, and Oscar Eggen, vice president of engineering.

But we in Charles City were so certain that Ford would release a more modern tractor that we devoted more time to the Super 55 design with the independent power take-off, helical-gear transmission, four-cylinder Super 66 engines, easy riding seat and many other innovations for a modern tractor.

Ford had a special sales show in Des Moines, Iowa, on November 30, 1953. Oliver's Des Moines sales branch had a combination Oliver and Ford dealer who invited me to attend the show. The show was the introduction of the Ford NAA Jubilee model. The Ford NAA had the advantages over the Model 8N that we had anticipated and that we were planning for the Super 55.

On the following Monday, there was a conference call between Oscar Eggen and Alva Phelps in Oliver's Chicago Office, and Plant Manager George Bird and me in Charles City. They gave us the approval to go ahead with the advanced Super 55 tractor design rather than our copy of the Ford 8N and requested a pilot run of 300 Super 55 tractors by June 1954. The Super 55 was to be introduced with the new Super 66, 77, and 88 in late summer. This gave us seven months to design, tool, and manufacture the pilot run.

Our Charles City engineering department had always worked closely with the foundry, tool design, machine shop and other related departments. We had some crash programs before, but this one was even more intense. We released the rear main frame drawings to start making production patterns before we had assembled a prototype. We kept in touch with the foundry and requested that they tell us the cut-off date after which no more changes could be made.

There was much give and take among all departments who shared the same goal of a pilot run of 300 in June 1954. We assembled four experimental models of which one was to remain in Charles City, Iowa, and three were sent to Bakersfield, California, for test under the supervision of our Field Test Engineer Joe Roland. All of the reports were good.

We conducted the usual accelerated tests on each component assembly, such as hydraulic system, transmission, and power take-off in the Charles City Experimental Engineering Department. The purpose of these tests was to prove the unit and compare the results with those from tests on assemblies with known components. We made the pilot run as scheduled in June 1954, and the tractors were distributed. Then, early production runs of 4,000 were scheduled to be manufactured in the fall of 1954. The first production run was completed and the tractors were shipped, when a call came from Joe Roland that a fan on one of the Super 55 tractors had broken and gone through the radiator.

This 1954 advertisement for the Super 55 touts Oliver's new model. The Super 55 was an entirely new utility tractor designed for future needs, including operator comfort and safety, easy maneuverability, and high efficiency. *Floyd County Historical Society*

After an intense crash program, the new Super 55s were first shown publicly when many Super 55s from the pilot run brought up the rear of the 1954 Charles City Fourth of July Parade. This Super 55 with a manure spreader demonstrates the high utility of this tractor. Its four-cylinder engine with aluminum pistons could attain speeds of 2,000 rpm. *Jeff Hackett*

We quickly reviewed and tested the Charles City tractor with a high-speed camera and a strobe light. We learned that the fan blade tips were bending about 3/16 inch at the tip when they passed by the lower tank of the radiator. This deflection of the fan blade was enough to cause an early fatigue failure. Generally, the fan blades do not pass by an obstruction such as the upper or lower radiator tanks, but in order to have a low-silhouette Super 55, this compromise was necessary. We set up a severe test quickly and failed a fan in less than four hours of operation.

Meanwhile, we were in touch with suppliers to obtain a more sturdy fan that would not deflect and fail. We obtained some samples to test and found one that did not deflect and did not fail in our severe test set-up. Our service department prepared a program to replace all Super 55 fans on the pilot and first production runs. I am not aware of any failure of the improved fan on any customer's tractor.

The Super 55 had helical gears in the first production assemblies. Because of the overlap of the gear teeth provided by the helical gears, the helical-gear transmissions were very quiet.

The Super 55 became the 550 in 1958. It was considered a top performer in its field. Its stability, easy riding seat, independent

PTO, helical-gear transmission, proven gasoline or diesel engine, and other advantages made it a formidable competitor. However, Oliver did not sell them in large enough quantities to lower the cost of manufacture to a competitive per unit cost. By this time, most other tractor companies had comparable utility tractors. In the early 1960s, less expensive foreign tractors began to dominate the small agricultural tractor market in the U.S. As coordinator of outside products from 1965 to 1970, I worked with Fiat in Italy to purchase tractors for Oliver—models 1250, 1450, 1255, and 1355.

Super 66, 77, 88, and 99

In 1953, Oliver Engineering at Charles City was asked to update the 66, 77, and 88 with more power and to introduce them as Super models. So, in addition to the crash program on the Super 55, we also had to provide for pilot runs of 300 for each of the Super 66,

The Oliver Super 66 tractor used the Fleetline Model 66 chassis with the Super 55 2,000-rpm engine with aluminum pistons. A Super 66 diesel tested at Nebraska in 1955 produced 22.49 drawbar and 33.69 belt horsepower. This is a 1954 Super 66 diesel tractor with a single front wheel. ***Jeff Hackett***

In the Super 77 tractors increased power was obtained from their six-cylinder engines by increasing the bore size from 3 5/16 inches to 3 ½ inches while using aluminum pistons and 1,600 rpm maximum speed. This 1955 Super 77 is a standard tread model. ***Jeff Hackett***

77, and 88 for introduction in late summer. At the same time we were preparing the paper on the XO-121 for the American Society of Agricultural Engineers summer meeting at the University of Minnesota in June 1954. We were also beginning the preliminary specifications for the experimental 1800 and 1900 tractors to be introduced in 1960. We were very busy!

The Super 66, 77, and 88 styling remained basically the same as the 66, 77, and 88 except the side panels to the engine compartment were opened to provide better movement of air and engine cooling.

The four-cylinder 66 engine was changed from a bore of 3 5/16 inches to 3 1/2 inches and the speed was increased from 1,600 to 2,000 rpm. The higher speed required counterbalancing the crankshaft and changing from cast-iron to aluminum pistons. Both the gasoline and diesel 66 engines were also used in the Super 55. The six-cylinder 77 engines were changed from a bore of 3 5/16 inches to 3 1/2 inches and the speed remained at 1,600 rpm. The 88

The Super 88 tractor design increased power output by increasing the bore size from 3 ½ inches to 3 ¾ inches with the same cast-iron pistons and a maximum engine speed of 1,600 rpm. A Super 88 diesel tested at Nebraska in 1954 produced 37.93 drawbar and 47.6 belt horsepower. *Jeff Hackett*

The Oliver Super 99 tractor was an updated version of the Fleetline 99 offered with an Oliver six-cylinder gasoline or with a 371 General Motors Detroit diesel engine. This gasoline version was built in 1955. *Jeff Hackett*

The three-cylinder two-cycle General Motors engines on the gasoline Super 99s were equipped with superchargers. ***Jeff Hackett***

engines, also six-cylinder, were changed from a bore of 3 1/2 inches to 3 3/4 inches and the speed remained at 1,600 rpm.

A pilot run of 300 each of the Super 66, 77, and 88 tractors was produced for introduction during the summer of 1954. Production runs of the Super models followed in the fall of 1954. The Super models continued in production through part of 1958.

The Super 99 was designed and built by Oliver's South Bend tractor plant. It had the basic chassis of the 99. The engine was the General Motors two-cycle 371 diesel, which had three cylinders with 71 cubic inches of displacement per cylinder. The different engine required a new front frame. The styling was the same as other Super models.

550, 660, 770 and 880

In 1957, Oliver Engineering at Charles City was asked to upgrade the Super models and to provide more power. The 550 and 660 engines had a bore increase from 3 1/2 inches to 3 5/8

This advertisement introduced the updated Super 55 to an all-purpose Oliver 550 tractor. ***Floyd County Historical Society***

inches and the speed remained at 2,000 rpm. The 550 continued in production from 1958-1975. The 660 continued in production from 1959 to 1964.

The 770 tractor engines were increased in speed from 1,600 to 1,750 rpm. Many other changes were made at this time, such as the crankshaft, aluminum pistons, and gasoline combustion chamber. The 770 continued in production from 1958 to 1970.

The 880 tractor engines had a speed increase from 1,600 to 1,750 rpm. The bore and stroke remained the same as the Super 88's. Other changes to the engines paralleled the 770. The 880 also had the helical gears, making these transmissions quiet. The 880 tractor continued in production from 1958-1963 when it was superseded by the 1600 model.

In 1956 Oliver hired an industrial designer, Wally Droegemueller. He was very helpful in the styling of the 550, 660, 770, and 880 and other tractor models and products in other Oliver plants. Wally continued in this capacity through the styling of the White tractors.

Power Booster Drive

The Power Booster Drive was introduced as an option for the Super 66, 77, and 88 models. It had gears, a housing, and an over-running clutch. The regular gears were direct drive to the transmission. When the direct drive was released, the over-running clutch had pawls that would grip a slower shaft and slow the speed of the tractor. The result was about 15 percent reduction in speed with approximately 15 percent extra pulling power. IHC was the first with this type of system which they called Torque Amplifier. Our goal was to provide a simple mechanism and reduce the need for power shift transmissions. Our Charles City engineering department continued to search for the ideal power shift transmission.

The model 550 and 660 tractors provided more power than the previous Super 55 and Super 66. This was accomplished by increasing the bore of the four-cylinder engine from 3 ½ to 3 5/8 inches while retaining the same maximum speed of 2,000 rpm. This is a 1961 model 660 with an adjustable wide front axle.
Jeff Hackett

Walt Roeming and Herb Morrell discuss the new 880 engine which included aluminum pistons and counterbalancing to allow for greater engine speeds.
T. Herbert Morrell collection

Carl Hecker, president of Oliver at that time, requested an experimental test be conducted with a General Motors four-cylinder diesel engine and a Hydramatic four-speed industrial power shift transmission. This combination was assembled in a Super 88 chassis which included an Oliver six-speed transmission. Oliver engineering was concerned about such a combination because of the cost and safety. The test was conducted in a field being plowed to plant corn. The field had a 30-foot wide waterway in the middle of it. Our test driver was informed as to what might happen when he actuated the Hydra-Lectric hydraulic system to raise the plow out of the ground. A fully loaded tractor with a plow could be unloaded in about one half of a second. The test revealed what we had anticipated. When the plow came out of the ground, the Hydramatic power shift transmission shifted quickly through its four speeds and the tractor's front wheels raised about four feet off the ground. The power shift transmission research for Oliver tractors continued.

A major change in the styling of Oliver tractors occurred with the introduction of the 550, 660, 770, and 880 models. The yellow trim was changed to off-white with a modernized grill. The new styling on this 1960 model 880 Standard tractor in the foreground contrasts with the older style of the 1955 Super 55 in the background. *Jeff Hackett*

The Power Booster Drive tests were completed in the laboratory and in the field with no problem. It was released for production. About this time, our suggestion committee decided to accept a suggestion to discontinue the transmission flushing operation on the assembly line. The flushing operation consisted of filling the transmission with oil, rotating the gears, and then draining the oil. The oil was then filtered and placed in a reservoir to be reused. When the transmissions were not flushed, sand, shop dirt, and other dirt from handling and machining would be in the transmission oil.

The Power Booster Drive got its lubricants from a system circulating through the transmission. The over-running clutch acted as a centrifuge to collect and retain all of these contaminant particles. Without the flushing operation, these particles built up

This model 880 has Wheatland fender/cover guards. An Oliver 880 burning gasoline was tested at Nebraska in 1958 and produced 42.89 drawbar and 57.43 belt horsepower. *Jeff Hackett*

and caused the surfaces of the over-running clutch pawls to wear. A worn over-running clutch would not grip the shaft and it became non-functional.

The suggestion committee generally consisted of a representative from each major department, although the representative from Engineering was not present when the decision was made to discontinue the flushing operation. The net saving from discontinuing the flushing operation was less than $2 per transmission, but cleaning up all of the tractors and replacing the over-running clutch and other worn parts on tractors that had been assembled and shipped cost millions of dollars. Guess who got blamed for the problem! Engineering. After that all transmissions were turned upside down, and about 40 streams of oil under pressure washed out the transmission just before the transmission openings were covered. A special lubricating system with a filter was included for the tractors having the Power Booster Drive.

The Oliver Model 950 tractor had a 302-cubic-inch engine with the bore and stroke both 4 inches and a maximum speed of 1,800 rpm. An Oliver 950 diesel tested at Nebraska in 1958 produced 48.76 drawbar and 67.23 belt horsepower. This is a 1961 Oliver 950 standard tread tractor that burns gasoline. ***Jeff Hackett***

950, 990, and 995

The 950, 990, and 995 were South Bend tractor plant products that were moved to Charles City in 1958. The 950 had six-cylinder Oliver diesel or gasoline engine with a bore and stroke of four by four inches The displacement was 302 cubic inches and the speed was 1800 rpm.

The 990 had the GM 371 diesel engine, but was otherwise basically the same as the Super 99. The engine speed was 2,000 rpm. The 995 was the same as the 990 except that it had a supercharger that was called the Oliver GM Lugmatic Torque Converter.

One-Row Offset Tractors: Super 44 and 440

Oliver had several programs to develop a one-row offset tractor to compete with the Farmall A, which had been introduced by International Harvester Co. in 1939. The small size of the little

The Oliver model 990 tractor had the same General Motors engine as the Super 99, the GM 371. An Oliver 990 Industrial tested at Nebraska in 1958 produced 61.46 drawbar and 75.46 belt/PTO horsepower. *Jeff Hackett*

Farmall made it popular with small-acreage farms, while the offset design made it an outstanding cultivating tractor. Specific aspects were patented, which made it difficult for other companies trying to provide a comparable tractor.

Some Oliver executives suggested that Oliver make a tractor with the transmission on the right side by the right rear wheel and angle the engine and drive line toward the left front. A one-row cultivator could be mounted on the center of the tractor's rear end. Several evaluations were conducted on the feasibility of such a tractor, but our negative evaluations were not understood by the persons originally proposing the tractor.

Oliver engineering at Charles City purchased an engine, transmission, and other assemblies to build one prototype tractor. The frame of the tractor was designed to fit these purchased items. The prototype tractor was referred to as "a tractor built to prove that a tractor shouldn't be built that way."

The Oliver Super 44 was designed as a one-row offset tractor for garden use or one-row crops such as tobacco. Most of these tractors were sold in the Carolinas and Virginia to cultivate tobacco.
Jeff Hackett

When I started to work for Oliver on October 4, 1944, the supervisor of experimental engineering, Tommy Martin, had a small farm a few miles south of Charles City where his close friends and members of the Oliver engineering department had victory gardens (home gardens planted to increase food production during World War II). This offset prototype tractor was available to all gardeners to use to cultivate their gardens, providing they were planted in 40-inch rows. The operator of this tractor could do very well for a short time while cultivating, but soon became hypnotized looking between his feet where plants were passing through a small opening behind the angled drive line and the rear of the tractor. When cultivating, one also needs to look up at the row for some distance ahead, but obstructions to this line of sight made it hard for the operator to keep the cultivator accurately on the row of plants.

One of my assignments in the late 1940s was to administer the one-row offset tractor design. The design we developed had the engine and transmission on the left side of the tractor. A prototype was built for viewing by some Oliver executives, including Alva

Phelps, Oliver CEO, and Oscar Eggen, vice president of engineering. We worked very late two nights before their visit, but there were a few finishing touches left to do the morning before the viewing by the executives. A starter had been selected and installed, but none of us had checked the direction of rotation needed for our tractor's special starter mounting. During the viewing, the starter turned the engine backwards, so we had to hand cranked it. This was one of my most embarrassing moments.

We operated this tractor at the plant and in the field. Then it was decided to build four more prototypes—known as model G experimentals—and send them to garden country for tests. These four units were loaned to a large farmer in the Rio Grande Valley in south Texas where we could get operation 24 hours per day. Herb Pyle was sent with the units to monitor their operation and send us his weekly report on each tractor.

We thought that we had designed a foolproof transmission interlock mechanism to prevent the transmission from getting locked into two gears, but, as in the original Fleetline testing, the operators managed to do it. Herb Pyle figured out how it could be done. The gear shift lever had to be pulled upward against the centering springs so that the end of the lever could pass over the notch in one transmission shifter rod while the tractor was in one gear. The gear shift lever then could engage the notch in another shifter rod. When the second shifter rod was moved to engage the gears for a new speed, the transmission locked in the two gears. The transmission cover would have to be removed and both shifter rods returned to their neutral position before the tractor could continue in operation.

In the case of the model G, the solution to the problem was different than in the 66, 77, and 88. We made some new transmission cover castings, machining the centering springs shallower and increasing their size. We sent them to Herb who changed the tractors and they had no more problems with the transmissions getting locked in two gears.

A market survey was taken to determine the number of Super 44 tractors the Oliver dealers could sell. As the result of the survey, the project was again deferred for more study and consideration.

The model 440 tractor is an updated version of the offset Super 44. This is the rear view of a 1960 model 440.

In 1953 the Charles City engineering department was asked to consider the possibility of modifying a Model 66 tractor to compete with the IHC Super AV tractor. The Super AV had higher crop clearance for cultivating tall crops, such as asparagus and other raised-bed crops. We modified the 66 in an attempt to provide a suitable tractor, making 250–500 tractors per year without the high cost of special tooling and small production runs.

The modified 66 tractor had a single-row cultivator mounted on the right side, 42-inch rear tires, increased front axle height, and other related changes. One unit was assembled with a mechanical power lift. One concern we had was whether the side draft cultivating only one row would be satisfactory. Our tests at Charles City were satisfactory, but the list price of this combination was approximately $150, or about 10 percent more than the IHC Super AV, so the proposal was tabled. Besides, the 66 itself had the power and flexibility to provide for two-row application.

In late 1953 or 1954, Oliver's Battle Creek, Michigan, aviation plant was in the process of discontinuing the Boeing Aircraft Co.'s fuselages for their reconnaissance plane, so Oliver was looking for products to continue plant operation. The Charles City engineering department was asked to revive the one-row offset tractor designs and send all drawings and information to Battle Creek No. 2 plant. Charles City engineering recommended using the new Cessna hydraulic system that Minneapolis-Moline was using at the time, or the Super 55 hydraulics, if they could be developed quickly enough.

The one-row offset tractor project did not stay at Battle Creek very long. It was transferred to Oliver's South Bend tractor plant. Their engineering department adapted the Cessna hydraulic system and released the model Super 44 in 1957. When the South Bend plant was closed in 1958, all tractor models were transferred to Charles City.

The Charles City Service Department soon learned that there were problems with the adapted Minneapolis-Moline hydraulic system. This hydraulic unit had many adjustment provisions which resulted in looseness between a large number of the control parts. The looseness in the control parts caused some erratic functions of ground-working equipment, such as a three-point hitch-mounted plow. We retained the most sensitive adjustment and then redesigned it to eliminate approximately 125 parts. We also attached the hydraulic unit more securely to the tractor. The Charles City engineering department then proceeded to adapt the 550 hydraulic system to the one-row offset tractor, which became the model 440 in 1959 or early 1960. Sales of the 440 were very low. Further consideration of adapting the modified 550 hydraulic system to the 440 was discontinued because the cost of the casting equipment and the tooling could not be justified economically.

CHAPTER FIVE

Another New Line: 1600, 1750, 1800, and 1900 Tractors

During the early 1950s the Charles City engineering department's research for the XO-121 and our experience with the Super models revealed some interesting economical issues. The main economical issues for farmers were the increased cost of maintaining an extra hired man, the availability of hired men, and the decline in the number of farmers. When we plotted these trends, it became obvious that by 1960, farmers would need extra power to farm the land that previously was farmed with the help of hired workers.

The 1800 was designed to be that Row Crop tractor to replace the hired hand. The 1900 was designed to be the larger Wheatland tractor. The 1600 and 1750 were smaller versions of the 1800 and were designed to be released later. The 1800 was used for most of the field and laboratory tests. The drive line of the 1900 was designed to meet the power requirements of more than 100 PTO horsepower.

Design criteria established for the larger Row Crop tractors included the innovations of the Fleetline 66, 77, and 88 with increased power requirements, extra fuel capacity, and draft control. Four-wheel drive and front tires smaller than the rear tires were also included design criteria.

Draft Control

Draft control is a function of the hydraulic system and the tractor hitch to the implement. When extra pulling power (draft) is required, the hydraulic system is actuated and it lifts the implement a slight amount to reduce the required draft. The implement tends

In order to meet the anticipated future needs of farmers, the Oliver 1800/1900 series of tractors were designed for increased power, increased fuel capacity, and draft control. The 1800 and 1900 were produced from 1960 to 1964 when they were replaced with the 1850 and 1950. This shows an Oliver 1800 doing spring plowing. *J.C. Allen and Son*

to lower itself to the preset position of the control lever when there is less draft required.

The smaller tractors had a three-point hitch with the upper link being the draft control sensing device to actuate the tractor hydraulic system. The two-plow implement on the small tractors was ideal for the upper-link sensing device. The front bottom of the plow tends to rotate rearward about the hitch points between the lower links and the plow. This tendency for the plow to rotate exerts pressure on the upper link to control the draft. But when a third bottom is added, it tends to stabilize the front bottom, causing the upper link sensing to the hydraulic system to be less effective. Four or more bottom plows will cause the front bottom to be ineffective for good draft control. A study of the overall draft control system led to consideration of a two-point hitch with semi-mounted implements and draft sensing from the two lower links.

The South Bend engineering department was not too enthusiastic about this new system and questioned if enough sensing from the lower links could be obtained. Engineering at Charles City did numerous mathematical calculations and obtained

Draft control was one of the most important features of the new series of models. Here Bob Johnstone of the South Bend plant explains the two-point draft control hitch on the five-bottom semi-mounted plow to three dealers. ***Chuck Dillman***

a four-bottom plow to convert from a pull type to a two-link semi-mounted plow. The furrow wheel was maintained to support the rear end of the plow. The first tests of the combined tractor and implement worked well, but needed some refinements. Oliver provided the two links with a mechanism to permit the operator to stay on the tractor seat while hooking up an implement designed for the new lower links.

Independent Power Take-Off

The independent power take-off (PTO) on the 1800 and 1900 had an oil clutch with hydraulic engagement inside the final drive area. There were several reasons for this location. The PTO had to be within the standards and there was not enough room for the clutch and the mechanisms at rear of the tractor. Also, the PTO clutch would have been too difficult to operate manually from the operator's seat.

In the late 1950s, newly developed implements began to require a higher PTO speed than the standard 540 rpm, so the new standard of 1,000 rpm was established. An example is this PTO-driven sickle mower on the 1966 model 1650. *Jeff Hackett*

Engines

The 1800 engine was a revised 880 engine with larger cylinders that could attain higher speeds. The 1900 engine was a two-cycle four-cylinder General Motors engine with 54 cubic inches of displacement in each cylinder. This GM 454 diesel engine was a relatively new design.

Cast-Iron Grill

Oliver's design criteria included the provision that one-third of the weight of the basic tractor rest on the front wheels and two-thirds of the weight on the rear wheels. We wanted the additional weight to be on the front of the front frame, so we added a heavy cast-iron grill to the 1800 and 1900 tractors. This added weight was necessary to balance the weight of implements added to equip the

Gasoline and diesel versions of the model 1600, both equipped with Hydra-Power, were tested at Nebraska in 1963. The diesel tractor produced 57.95 PTO and 46.9 drawbar horsepower. The gasoline tractor put out 56.5 PTO and 47.09 drawbar horsepower. ***T. Herbert Morrell collection***

tractors for field use, and was crucial to the overall stability of the 1800 and 1900 tractors.

Steering

These larger tractors were difficult to steer manually. First, the universal joint angles exceeded their angle capacity. Second, without some assistance, steering of the larger tractors in most conditions of operation required too much effort for the operator.

The Char-Lynn company had just developed a hydrostatic steering mechanism, which permitted us to provide a tilt and telescoping steering wheel. With the tilt and telescoping of the steering wheel, the operator could stand up to rest from sitting and adjust the steering wheel to operate it comfortably when standing. In 1970 at the Farm Progress Show in Iowa, J. I. Case had an area for demonstrations and a program on what was new on their tractors and equipment. One item on the program was about the "new" tilt and telescoping steering wheel. One bystander said to another bystander, "Oliver has had the tilt and telescoping steering wheel for about ten years, haven't they?"

Fuel Capacity

Oliver engineering listened to comments from everyone during the development of these new tractors, and having enough fuel capacity for an eight-hour day was a prime concern. Oliver developed additional fuel capacity by designing wheel guard fuel tanks as supplementary to the main fuel tank. Finding space for everything in the design of the new tractor model was a problem. A familiar saying among the engineers was, "Let me hang the fuel tank and the battery compartment on a sky-hook and then I will have enough room for everything else."

Four-Wheel Drive with Terra Tires

The 1800 and 1900 four-wheel-drive tractors became a natural for Terra Tires made by Goodyear. The Terra Tires required very low pressure and the ground area covered by the Terra Tires was much larger than for the regular tractor tires because the bars on the Terra Tires were much lower than on the regular tractor tires. The lower bars with the lower pressure gave greater contact with the ground, which made this combination suitable for special requirements At one of the Farm Progress Shows in Iowa, Oliver representatives, guided by Dutch Zandbergen, pressed chicken eggs with small end down into the wet ground and then the tractor with Terra Tires was driven over the eggs without breaking the eggs. The low ground pressure was advantageous when the tractors were used in fields growing sod and other crops requiring that no tracks be left.

When Oliver engineering was researching the applications most suited for Terra Tires, I arranged to get a loan of a 1900 with Terra Tires to be used for a short time in the sugar cane fields south of Lake Okeechobee, Florida, where crawler tractors were being used. This arrangement was satisfactory to Aubry Hedrick, the manager of the Atlanta, Georgia sales branch. A few weeks later I contacted Aubry to find out how the unit was performing. His reply was that he had sold it to a sod grower, a minor potential use of this special tractor. So much for an attempt to assist the Oliver sales department.

Roll Over Protective Structure

Oliver emphasized safety considerations in the design criteria for the 1800 and 1900 tractors. The operator was to enter from the left in front of the rear wheel instead of from the rear. The axle carriers were to be designed with enough strength to avoid failure when a fully equipped tractor was subjected to a roll-over accident equivalent to twice the force of gravity. Oliver engineering became interested in Roll Over Protective Structures (ROPS) as a result of accident reports and a 1953 Swedish technical paper, so ROPS were included in the design criteria.

When the 1800 and 1900 tractors were introduced in 1960, ROPS air-conditioned cabs and ROPS canopies were presented for production. However, a market survey by the Oliver sales department indicated that so few would be sold that the cost of tooling would not be recovered within the required time limit. Thus, the introduction of ROPS was deferred until 1969 when ROPS became accepted by customers. In the meantime, Oliver engineering participated in the development of the ROPS standard. The specifications became more detailed and complete when the ROPS standard was developed and printed in 1968.

Testing

The first field tests were on farms near Charles City. One experimental model was tested near Lubbock, Texas. There were tests on the 1800 & 1900 near Clarksdale, Mississippi. Neither of these areas were good remote test areas because the number of hours of operation within one year was quite limited. Some tests were at Green Giant near Le Sueur, Minnesota, in the summer. Green Giant used the tractors almost 24 hours per day. They often had to plow in the rain and with water running in the furrow to keep up with their farm schedules. They used one 1900 tractor 1,000 hours during a 1,200-hour period.

The experimental 1800 and 1900 tractors had two major and one minor redesign which resulted from the field and laboratory tests. We received many helpful comments from the users of the experimental tractors during the years of 1955 to 1959.

Setbacks

The 1800 and 1900 development program had some setbacks. Our development budget was cut because 1956 was not a good sales year, so we could not make any more experimental parts. We could only do design work and follow the field tests of the experimental tractors already built. The Oliver sales department in our Chicago office was not in favor of the 1800 and 1900 program. They said that these new tractors were wrong, they cost too much, and, according to a market survey, they could not be sold in large quantities. Then, 1958 was another year of low sales and we had to cut our budget again. However, we could redesign and test updated experimental models.

During the middle of 1958, we were approaching the release of all of the drawings for the 1800 and 1900 tractor production. One morning Oliver CEO Alva Phelps and President Carl Hecker flew by company plane from Chicago to Charles City. Plant Manager George Bird picked them up at the airport and all three came to the engineering conference room. Carl Hecker said, "This meeting won't take long. We are cancelling the X89 or 1800 and 1900 program."

George Bird's answer was, "Before we make a final decision, let's go out and see an 1800 plowing in comparison with the latest J. I. Case tractor and plow." The 1800 was plowing about 25 percent faster with one more plow bottom than the J. I. Case tractor and plow. George Bird asked Phelps to ride with him in his air-conditioned Cadillac. They followed the two tractors for about thirty minutes. When they returned to the end of the field, Phelps got out of the car and said to Hecker, "We can't cancel this program because it could be Oliver's salvation."

Introduction of the 1800 Series

The 1800 and 1900 tractors were introduced in November 1959 at the Hippodrome in Waterloo, Iowa. All of the dealers were flown to Waterloo in groups, each group for a two-day meeting. The sales department had changed its attitude regarding the merits of these new tractors. This introduction of the 1800 and 1900 tractors and implements was a very successful presentation.

Herb Morrell is shown driving the 1900 with a plow into the arena where the dealers viewed the new tractor for the first time. He then presented the specifications and advantages for both the 1800 and 1900 tractor. *Chuck Dillman*

In November 1959, Oliver held a show at the Hippodrome in Waterloo, Iowa, to introduce the model 1800 and 1900 tractors. Dealers were flown in for this event. Model 770 tractors were fitted with power steering and Funk Reversomatic transmissions and were demonstrated in a tractor square dance driven by women. This was intended to demonstrate how easy the tractors were to operate, one of Oliver's design criteria for these new tractors.

Since 1960 was an election year, the theme "Vote for Power" was used to promote Oliver tractors. This was a "demonstration" in the Hippodrome to emphasize this theme for the dealers. The dealers were then invited to view Oliver's new products. ***Chuck Dillman***

The opening feature of the show was a square dance by eight Charles City housewives driving tractors that were equipped with power steering, our regular six-speed tractor transmission, and the Funk Reversomatic between the flywheel and the transmission, which allowed the tractors to be shifted into any one of their speeds. A separate lever was used to reverse to the direction of motion. This allowed easy maneuverability, which was dramatically demonstrated by the beautiful young women dressed in western style in Oliver colors driving the tractors in a square dance.

"Most vigorous applause of the evening was given the feminine drivers, costumed in western style and Oliver colors. The women have been practicing the driving patterns for almost three months," wrote the *Charles City Press* on November 12, 1959.

The 1800 and 1900 tractors were then driven into the arena by Walt Gardner and me. After our presentations, there was a demonstration of Oliver personnel campaigning to elect these two new model tractors into the Oliver line in the Vote for Power campaign. Then the dealers were invited to view the new 1800 and 1900 tractors and updated products from the other Oliver plants.

Both the 1800 and 1900 tractors were available with Terra tires. These large, low-pressure tires distributed the tractor's weight over a larger area and, therefore, decreased the weight per square inch tremendously. These tractors could be used where tractor tracks were undesirable, such as in sod production.
T. Herbert Morrell collection

We announced at the meeting that these new tractors would be available for sale by mid 1960. The production run was increased three times by mid 1961 and the quantity of four-wheel-drive tractors sold exceeded everyone's expectations.

Further Model Upgrades

The frequent upgrading of the tractors and change of model numbers continued through the 1550, 1650, 1750, 1850, and 1950. The next upgrading and model change became the 1555, 1655, 1755, 1855, and 1865.

The versatility of the 1800 and 1900 continued from the Fleetline tractors. When the 1600 was introduced in 1962, it became very popular. Not more than six months later, I received requests for approval of two special tractors in one day. A visit with Morrie Thelen of our specifications department revealed that we had approved 1,172 different 1600 tractors. I asked him to tell me how many tractors could be assembled without designing a new part. He came back with the answer of more than 2,600,000. Oliver had become a truly custom-built tractor company.

Certified Horsepower

During the early 1960s, the Oliver tractor plant was criticized because the horsepower quoted for our tractors was not as high as some competing tractors of the similar size. The difference was that Oliver published the horsepower of the actual production tractors while some competitors published only horsepower determined by the Nebraska Test results. The Nebraska test tractor was generally tuned to provide the absolute maximum horsepower which would be higher than most tractors could produce under normal operation by the customer.

To provide information to Oliver salesmen and dealers, a Certified Horsepower decal was placed on the sheet metal on the left side of the tractor. This decal gave the horsepower of that particular tractor when it was assembled and tested. Tractors usually produce greater horsepower after they are broken in than at the time

The 1950 was offered in two-or four-wheel drive. This 1965 model 1950 with a General Motors 443 diesel engine produced 105-79 TPO horsepower. *Jeff Hackett*

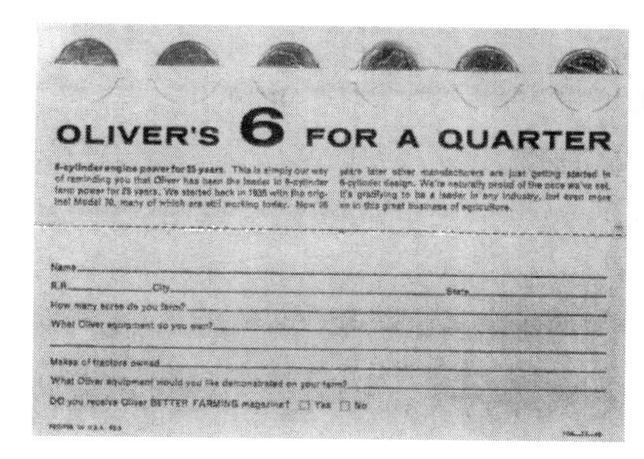

OLIVER'S 6 FOR A QUARTER

6-cylinder engine power for 25 years. This is simply our way of reminding you that Oliver has been the leader in 6-cylinder farm power for 25 years. We started back in 1935 with the original Model 70, many of which are still working today. Now 25 years later other manufacturers are just getting started in 6-cylinder design. We're naturally proud of the pace we've set. It's gratifying to be a leader in any industry, but even more so in this great business of agriculture.

Name ______

R.R. ______ City ______ State ______

How many acres do you farm? ______

What Oliver equipment do you own? ______

Makes of tractors owned ______

What Oliver equipment would you like demonstrated on your farm? ______

Do you receive Oliver BETTER FARMING magazine? ☐ Yes ☐ No

The 6 for a Quarter promotion was introduced to emphasize that, in 1960, Oliver had been building six-cylinder tractors for a quarter of a century. This Oliver advertising campaign countered John Deere's introduction of their six-cylinder tractors. *Robert Tallman collection*

of assembly. The Certified Horsepower decal related that this tractor would produce the listed horsepower or more when broken in.

6 for a Quarter

Oliver first introduced smooth-running six-cylinder engines in 1935. International Harvester introduced a new agricultural tractor in 1960 at the Ohio State Fair in Columbus, Ohio. A large banner at the display featured "Six-Cylinder Smooth Power." John Deere

The 1900 series were Wheatland tractors with increased power to drive more than 100 horsepower implements. The rollover protective structure was designed to be introduced in 1960, but was deferred until later when it was accepted by customers. *Jeff Hackett*

You drive in complete comfort, seated or standing, and no matter what the weather

Sturdy cabs with extra room inside, with full visibility on every side

Contour-tailored Oliver Continental cabs—Row Crop and Wheatland types—offer extra room, comfort and visibility all around. Pressurizer with overhead blower brings in 400 cubic feet of fresh air every minute. Protected, long-life, pleated-paper filter beneath the visor is easy to service. Sun-reflecting roof is insulated with fireproof, closed-cell polystyrene. Tinted safety glass, padded armrest and exhaust extension are standard. Optional air conditioner, heater, windshield wiper, outside mirror, radio and stereo-tape player.

The welded, box-section cab frame is 400 percent stronger than a hot-rolled angle iron structure. Continental Row Crop cab has man-shaped, outward folding doors and a mounting step on each side. Wheatland cab is equipped with Dutch-style rear door and casement-type side windows.

Still another type is the narrow-row Farmer's cab. It permits a wheel tread of 60 inches and is equipped with pin-mounted doors and windows. Pressurizer brings in 500 cubic feet of fresh air every minute. Optional wheel guard fuel tanks holding a total of 30 gallons are available.

Low-cost Converta-Cab and Weather Break also offered

A new way to take the sting out of the weather is with the Converta-Cab, a light weight, low cost unit which your dealer is prepared to install. Cab consists of wing-type doors, windshield of tinted safety glass, clear vinyl side windows and a fabric curtain of nylon-reinforced vinyl. Doors, windows and engine panels can be removed quickly, leaving just the sunshade for "fair weather" operation.

Economical canvas Weather Break also available.

Hydrostatic power steering with a tilt-telescope wheel

You can assume any driving stance or position on the spacious platform of a 1750 or 1850. Steering column tilts to three positions, makes stand-up driving restful and safe. And, the recessed wheel telescopes five inches to suit arm length.

Weigh yourself in, settle down in a contoured seat

Here's a seat with comfort control. A special load indicator tells you when you have adjusted the tension of a steel torsion spring to match your exact weight.

A deep-cushion armrest permits you to sit side-saddle without interference while you watch trailing implements. There's a wide range of fore and aft adjustments to suit leg length, whether you're 6-foot-5 or 5-foot-6. Backrest is adjustable in height, and a high, auxiliary backrest is available.

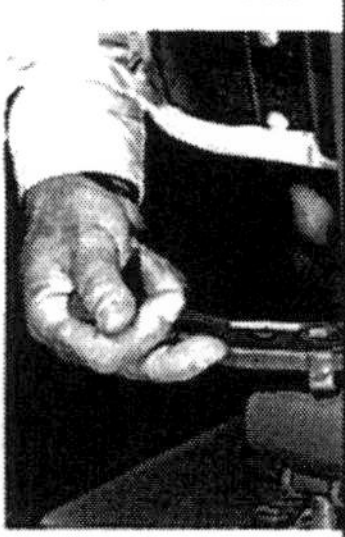

OLIVER CORPORATION, CHICAGO, ILLINOIS 60606

Subsidiary of White Motor Corporation

This advertisement from 1969 touts features found on the 1750 and 1850. The tilting and telescoping steering wheel, comfortable seat, and cab options added up to increased operator comfort.

Floyd County Historical Society

During 1967, the Oliver sales department suggested the development of an over and under auxiliary drive known as Hyraul-Shift. Coupled with the standard six-speed transmission, the feature essentially gave the tractor 18 forward speeds. This feature was offered on the 1650 through 1850 tractors. ***J.C. Allen and Son***

also introduced their new six-cylinder 4010 the same year. The inference was that this was something new in agricultural tractors, but Oliver had introduced five different models with six-cylinder engines during the 25-year period. Oliver's counter advertising program, "6 for a Quarter," consisted of 12,000 folders with cut-outs for six nickels that were sold for a quarter (25 cents). This meant "six-cylinder engines for a quarter of a century."

Over and Under Hydraul-Shift

During the late 1960s, the Oliver sales department suggested the development of an over and under auxiliary drive. The basic tractor transmission could be shifted into any of six speeds. The control lever on the Hydraul-Shift could be moved to over-drive in any of the six speeds, which would increase the tractor speed by approximately 15 percent and reduce the pulling power by 15 percent. The control could also be shifted into under-drive to reduce the speed and increase the power by 15 percent. The effect of this feature was to give the tractor 18 forward speeds.

In 1960, Oliver had been building six-cylinder tractors for 25 years. The 1935 Model 70, the first six-cylinder tractor is shown with the 1960 Oliver 1800 and Charles City plan personnel (from left) Bob Burgraff, Walt Gardner, Bill Sheely, Jim Light, Herb Morrell, John Dorwin, Gene Brunsman Jr., Merle Hicks, and George Bird. ***Carl Rabe***

Implements

The implement plants were also innovative and made contributions to the agricultural industry and society. The Oliver South Bend plant developed the throwaway Raydex Shares, a straight-nosed, simple design of a plowshare. Other plowshares, which had a long snout that was easily bent or broken, were more complicated and costly to maintain. In contrast, the Raydex shares were disposable, which lowered the cost of maintaining the plows. Oliver also made them for Ford.

The Shelbyville plant developed and patented an automatic bale thrower that threw bales of hay into a wagon without the aid of a second person to store the bales. A license agreement to make the bale thrower was issued to at least one other company.

CHAPTER SIX

Special Tractors

Oliver 25 Airport Tractor

Even before the introduction of the Fleetline, Oliver tractors had special uses. For example, the Oliver 25 was a model 70 Standard tractor with special tires and a special drawbar and hitch for towing small planes, cargo, and other airport materials. This tractor was used by a large number of airports during World War II. When larger planes were developed after World War II, larger and more special equipment was needed to perform the same functions.

The versatility of the Oliver 70 tractor made it easy to customize for special applications.

Military Standardized Engines

After the Korean War in the early 1950s, the tractor industry was considering the standardization of engine parts subject to high wear and early replacement, such as connecting rod bearings, main bearings, valves, and other related parts. Waukesha Motors and Oliver participated in the standardization program for gasoline engines. At that time there was a demand for a large number of tractors to perform maintenance in U.S. military installations around the world. From 1953 to 1955, Oliver received some of these tractor orders which specified the military standard engine. Some competitors took exception to the military standard engine and bid tractors at a much lower price. Oliver bid on some proposals with the military standard engine and with the regular commercial tractor engine. Oliver made little or no profit for all of its efforts to provide military tractors.

Alcohol Tractors

During the early 1950s, the country of Formosa (now Taiwan) wanted tractors that would burn pure grain alcohol that they could

The Oliver 66 Military tractor was equipped with the required standard engine parts. This tractor with a hydraulic mower was used for mowing in and around military installations. ***Carl Rabe***

produce from sugar cane. Gasoline was very difficult for them to get and very expensive. Some previous tests had been conducted using grain alcohol with Oliver's 77 gasoline engine at Waukesha Motors, in Waukesha, Wisconsin, so we started a program to use the same carburetor with other modifications on an Oliver 77 tractor. Meanwhile, Oliver sent a bid to Formosa and was selected for the contract. We learned quickly that there were great restrictions on obtaining pure grain alcohol in the U.S. We filled out many forms and wrote letters about our use of the alcohol and how we would control it. By the time we got through all the red tape required to get alcohol for testing, Formosa had solved its problem and was able to get gasoline at a reasonable price, so the Formosa project was canceled.

Special 66 Diesels for Banana Fields

We received an inquiry from a banana company in Guatemala asking if we could provide 50 diesel tractors that had no electrical system. They were having trouble maintaining the batteries and were losing gasoline from their gasoline tractors in the humid, tropical climate.

One cannot hand crank a diesel tractor unless it has some kind of compression release. No type of hand cranking of even a small diesel has been satisfactory because of the high compression ratio. We located a source that manufactured hydraulic starters and designed a system that included a hand pump and an accumulator to store a volume of hydraulic fluid at high pressures. Approximately 50 strokes of the hand lever were required to build enough fluid and pressure in the accumulator to rotate the engine at the rate of 450 revolutions per minute for one to three revolutions. This was enough to start the 66 diesel tractors under their tropical conditions.

The tractors were used primarily for pulling banana carts from the fields of banana trees. The banana companies liked the units, but Oliver did not get any repeat orders because the workers objected to the hand pump.

Remote-Controlled Tractors

On the Morrell farm in eastern Kansas we had an Avery tractor built in Peoria, Illinois. It was purchased around 1914 for belt

Banana plantations were experiencing problems maintaining the batteries and other electrical devices in the tropical climate, so they asked Oliver to develop a tractor with no electrical system. This shows a 66 tractor with a unique, non-electric, hydraulic starter. The long lever was used to manually recharge the hydraulic oil accumulator. *Carl Rabe*

work to power a rock crusher since our family was in the concrete construction business. This tractor had two speeds forward, 1.75 and 2.67 miles per hour, and it could pull a two-bottom plow. This tractor was used for many years to build silos, chicken houses, water reservoirs, and barns. The tractor wore out and it was difficult to get repair parts. A new Avery tractor was purchased about 1920 and the old one was kept for parts.

In 1924 my brother Paul put a furrow guide on the newer Avery tractor. The furrow guide consisted of a wheel in front of the tractor's right front wheel that was connected to the tractor steering mechanism. The guide wheel would follow the furrow and guide the tractor. Our automobiles consisted of a Model T Ford touring and a Chevrolet Model 490 and we did all of our own service work on them.

One Saturday we took a car and a saddle horse along with the tractor and plow to the field. While the tractor was plowing without an operator, Paul was fixing the car. When the tractor neared the other end of the field, Paul would ride the horse to the tractor, turn it around and start it plowing back toward the car. He would then ride the horse back to the car. When the tractor neared our end of the field, Paul would turn it around and start it toward the other end of the field. Paul then could get more work done fixing the car. The tractor was operated in the lowest gear of 1.75 mph. It seemed that it would take forever to plow that field. My function was to watch the tractor and to hand tools to Paul when he asked for them. This was quite an experience for me at eight years of age.

The Avery tractor was started by putting a lever on the flywheel axle and then into a notch on the outer edge of the flywheel and pulling down on the lever. The tractor was very difficult to start when the temperature was lower than 45 degrees F. We had a very large draft horse called Mace that was not nervous around machinery. On some of these cold occasions, Paul wrapped a rope around the flywheel and tied the other end of the rope to a single tree hitched to the tugs of Mace's harness. Mace was then led and the rope rotated the flywheel. This was a great idea until one day the tractor backfired and dragged Mace rearward. After that incident, poor old Mace refused to pull the rope to start the tractor. Obviously, he did not want to be pulled backward by this contraption.

During the early 1950s, Oliver started a research project to control a tractor from a remote position. A graduate student at the University of Nebraska had started the project with radio controls. He came to Charles City and worked with Charlie Adams in our experimental engineering department to apply the radio controls to an Oliver 88 tractor. Using the remote radio control, the tractor engine could be started; transmission gears shifted; the engine speed increased and decreased; the clutch engaged and disengaged; the tractor steered; implements lifted and lowered; left, right or both brakes applied; the tractor could be stopped; and the ignition turned off.

The radio-controlled tractor really worked. It was a novel idea to sit at the end of a field and operate the tractor by remote control, but it had some disadvantages. The tractor had to remain within sight, making it impractical for large or hilly field operation or while precision cultivating. Still, we learned much from this research project.

Oliver Engineering never felt this type of remote control of an agricultural tractor was practical. Since that time there has been research on providing guide wires in the soil for the tractor to follow using radio control of the tractor. But to my knowledge there has been no further development of this system because it would be costly to install and maintain, and would have a very low payback.

Industrial Tractors with Hydraulic Mowers

The Industrial 66, 77, and 88 tractors with hydraulic mowers were other special tractor combinations. The mower was driven by a special hydraulic system driven by the tractor engine. The mower was very powerful and could cut small trees up to about an inch in diameter.

Several of us from Oliver visited an unusual area where the tractor with mower was demonstrated. This was on a mountain

Industrial tractors equipped with hydraulic mowers were used for special mowing situations. Their flexibility, rugged construction, and hydraulic drive provided good results. This is a Oliver Industrial model 550 with a hydraulic mower. ***C. J. Gibbs***

plateau near Archbald, Pennsylvania, northwest of Scranton. A tractor with a hydraulic mower had been loaned to the owner of the plateau to mow brush that was killing the wild blueberry plants. The owner had attached a wick-like set of ribbons that had a poison to kill the approximately six-inch high brush stumps. If this tractor and mower worked, the owner planned to buy about 20 industrial tractors with hydraulic mowers to mow the brush on several thousand acres of the plateau.

But the owner also had an apple processing plant that supplied apple slices to bakeries in New York City and other large cities in that area. He used a toxic chemical to keep the apple slices looking freshly cut while being transported from his plant. The last information I received was that the tractor and mower had been picked up and the owner of the plateau was in prison for his use of toxic chemicals.

Super 77s for Kansas Turnpike Authority

The Kansas Turnpike Authority was responsible for the construction and maintenance of the Kansas Turnpike and it needed tractors. To meet their specifications, Oliver worked with mower manufacturer Danco of Claremore, Oklahoma.

The Super 77 adjustable-front-axle tractor was chosen. The plans included six-foot-wide rotary mower units to be mid-mounted between the front and rear wheels and driven by the independent power take-off on the tractor. The mower had to meet certain specifications to prevent the throwing of debris such as beer bottles at passing vehicles. Many types of discharge material deterrents were tried to meet the Kansas Turnpike specifications. We finally put log chain links over the mower openings to prevent rocks, glass, and other debris from being thrown at passing vehicles. The final test included the mowing of some scrub oak type trees and bushes. Any tree that could be bent down by the tractor front axle had to be mowed down and reduced to wood chips.

The tractors were painted dark blue and bright yellow, as specified by the Kansas Turnpike Authority. Whenever we traveled the Kansas Turnpike, we saw these tractors in operation or parked on the right-of-way.

This Super 77 was equipped with a Danco rotary mower and chains to meet the Kansas Turnpike Authority's specifications. The mesh screens were added so the mowers would not throw rocks onto the passing vehicles. These bright yellow and dark blue tractors could be seen for many years alongside the Kansas Turnpike. ***Carl Rabe***

Numerous other companies adapted their products to Oliver's Fleetline tractors. Arps made a trencher and Sherman a backhoe, and there are no doubt many more. These special tractors helped to keep the Charles City tractor plant open during low sales of conventional agricultural and industrial tractors.

Oliver Power-Pak

During the mid to late 1950s Oliver produced the Power-Pak, which allowed other companies to use Oliver's versatility and excellent designs for their own applications. This unitized construction included a gasoline or diesel engine, clutch or Reverse-a-Torc transmission, differential, final drive, axles and hubs, double-disc brakes, steering gear with power steering option, instruments, controls, and seat. Then other companies could take advantage of Oliver's innovative designs in producing specialized equipment.

OLIVER

SPECIFICATIONS

POWER-PAK

A new money-saving idea for powering mobile equipment

UNITIZED CONSTRUCTION INCLUDES:

- Engine Horsepower ranges from 45 to 127 Engine H.P. (gasoline or diesel)
- Clutch—Reverse-o-Torc—or Hydra-Power Drive
- Transmission
- Differential
- Final Drive
- Axles & Hubs
- Differential Disc Brakes
- Steering Gear—manual or power steering
- Instruments & Controls
- Seat

Model 1950 Power-Pak

This "unitized" power package is complete with engine, controls, transmission, and final drive, ready to "drop in place" on your machine to greatly reduce your engineering and assembly costs.

Certain special requirements can be met and the complete assembly can be delivered to you without excess parts or surplus material thereby simplifying your engineering, purchasing, and production problems.

Oliver will work with your engineers to meet your requirements and apply these units to your machine models in the most economical way possible. A large number of applications are now being used by original equipment manufacturers.

If you are contemplating new models or changes in existing models, it will pay you to investigate an Oliver Power-Pak as a possible power source. Experienced engineering service is available to assist in adapting the Oliver Power-Paks to your specific requirements.

Model 1850 Power-Pak

Currently being used on . . .

Pneumatic-tired road rollers

Rotary broom sweepers

High lift loaders

Pulp-wood harvesters

Harvesting equipment

Other self-propelled equipment.

Model 770 Power-Pak

The Oliver Power-Pak was very popular for many applications. Most companies purchased engines, transmissions, and other parts from Oliver to assemble their own units. ***Floyd County Historical Society***

Pneumatic Road Roller Units

Oliver Power-Paks were used to power pneumatic road roller units that compacted the soil for road and highway and other such construction. Oliver worked with Ferguson of Dallas, Texas, and Tampo in San Antonio, Texas. Oliver provided the power and the propelling device in Oliver Power-Pak units prepared for this application, consisting of the basic tractor minus the front and rear wheels and tires. Special axles were provided so that the road roller could be driven by using sprockets and chains from the tractor axles. Other companies, such as Bros, manufactured road roller units that also used the Oliver Power-Pak.

Oliver offered their tractors without wheels as a Power-Pak for other manufacturers to adapt for special applications. This Super 66 Wheel Roller is based on an Oliver Power-Pak. It was still in use when this photo was taken in 1996, more than 40 years after it was built. *Jeff Hackett*

Tampo Pneumatic Road Roller units were built using the Oliver Power-Paks. On the left is a small unit with a gasoline engine. In the center is a larger unit with a diesel engine, and the large unit using a twin 990 Power-Pak is on the right. *Carl Rabe*

Twin 880 Tractors

The Euclid Division of General Motors had a crawler tractor that had two engines and two transmissions mounted on the same frame. Each unit drove one of the two tracks. The twin combination was very powerful as a bulldozer and doing other types of work needed to prepare the base for a highway. George Bird, our plant manager, was very intrigued with Euclid's twin unit. He always showed great interest in anything new or novel. He reasoned that Oliver could make a special unit with twice the power of the 880. We tried to discourage him from pursuing such a project without a market survey, but we agreed to submit a development project to build one unit and then evaluate the usage later.

In November 1958 I was scheduled to be gone for two weeks to check on the inventory of a large number of attachments for Oliver Industrial tractors that had been purchased for the Industrial sales department. Before I left, we had a staff meeting to establish what

The Twin 880 was another Oliver experiment. In theory, such a unit sounded promising. Oliver engineers traveled extensively, but they found no market for a twin unit. It was limited to pulling implements such as this sheep foot roller.
T. Herbert Morrell collection

was to be accomplished while I was gone. When I returned home and reviewed the status of each project, I found nothing had been accomplished on the assigned projects. Each project engineer had been reassigned by George Bird to the design of the twin 880. The result was that some crucial engineering projects had been delayed.

Six Twin 880 units were built. One Twin 880 unit was demonstrated at the Oliver International show in Libertyville, Illinois, in June 1959. Sam White Jr., president of Oliver International S.A. was the master of ceremonies of the program. He introduced George Bird who talked briefly about the Twin 880 unit. Foreign countries showed no interest in such a unit.

One Twin 880 was sold to the Santa Anita Race Track in California to remove the horse race starting gates before the horses came back around the track. Another unit was sold for pulling a sheep's foot roller used for compaction in highway construction. A third Twin 880 Power-Pak was used in a Tampco pneumatic road roller. The other Twin 880s were disassembled and the parts were used to assemble regular 880 tractors, which were sold as used tractors.

Crawler Tractors

During the early days of crawler tractor production, the Cletrac, made by the Cleveland Tractor Co. was rated among the best. The Cleveland Tractor Co. was begun in 1917 as the Cleveland Motor Plow Co. and was purchased by Oliver in 1944. The industrial sales were controlled by the Cletrac sales department. Their main emphasis for many years was on crawler tractors and attachments, such as front-mounted loaders. Cletrac models continued to be produced at the Cleveland plant until 1963 when that plant was closed and the crawler tractors were moved to Charles City, Iowa.

Lull Loaders

Oliver had a special working relationship with Lull Engineering of St. Paul, Minnesota, which became an original equipment manufacturer using Oliver's basic tractor units. Lull was one of the first to produce front-end loaders for Oliver's Industrial 80 tractor

By the time Oliver bought Cletrac in 1944, they already had an amazing line of crawler tractors from the 10 to 55-horsepower range. This example is a 1950 model HG. *Jeff Hackett*

Powered by a 895-cubic-inch Hercules six-cylinder engine and weighing in at more that 16.5 tons, this OC-18 produced 133 drawbar horsepower, which could pull more than 31,000 pounds. *Jeff Hackett*

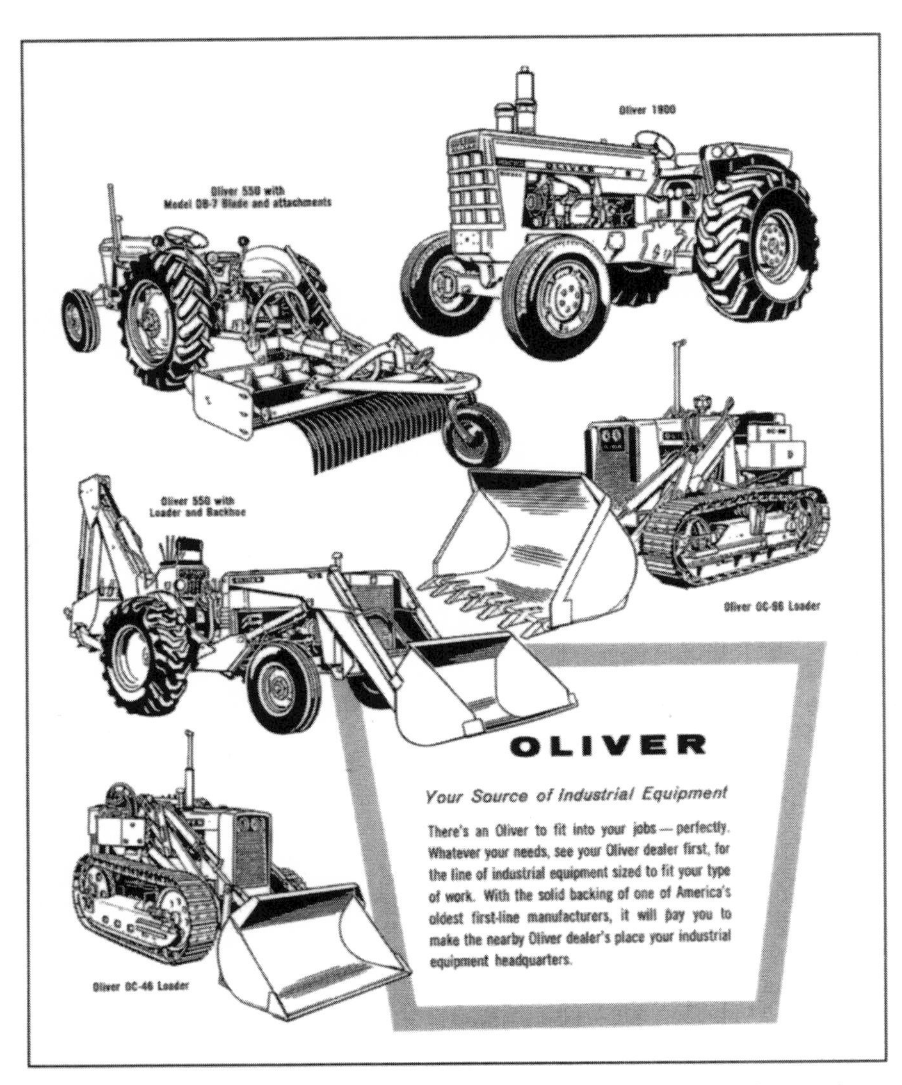

This advertisement shows a variety of Oliver models offered with loaders and other implements. "There's an Oliver to fit into your jobs–perfectly. Whatever your needs, see your Oliver dealer first, for the line of industrial equipment sized to fit your type of work." ***Bob Tallman collection***

This 1958 OC-4 crawler tractor and loader was still at work when this photo was taken in 1996. The OC-4 had a 130-cubic-inch Hercules three-cylinder engine available in gasoline or diesel configurations. *Jeff Hackett*

One application of Oliver Industrial tractors was for road work. This road grader was based on the Oliver 50. *J. C. Allen and Son*

Lull's model 7C used a 770 Power-Pak. These were sold in reasonable quantities for two or three years. The unit was valuable for constructing houses or other low buildings. ***Glenn B. Bazen/Lull Collection***

and then made a front-end loader for early Oliver 88 Industrial tractors.

Lull also made a high-lift loader used to place construction material on the second and third floors of a building under construction. Their first high-lift loader, the model 58, was built around an Oliver 66 tractor. The tractor provided the power and the drive mechanism. Lull designed a frame to accept the modified 66 tractor. Special hydraulics, lift mechanism, and other such assemblies were required by Lull to complete the vehicle. The model 58 High-Lift Loader had the capacity to lift 2,000 pounds to a height of 22.2 feet. This unit was very successful for small construction jobs, but there was a need for higher lifts.

Later Lull's model 7C used an Oliver 770 engine, transmission, and final drive. It had the capacity to lift 3,000 pounds to a height of 40 feet. Since the 7C, Lull high-lift loaders have developed into much larger and higher capacity units. Lull then began to

For tall, commercial construction, Lull designed a heavy high-lift loader with four-wheel drive. This unit was upgraded many times to meet the requirements of larger construction sites. *Glenn B. Bazen/ Lull Collection*

The Lull Loader on a Super 88 tractor is typical of the front-end loaders for the Industrial 77 and 88 tractor series. These front-end loaders provided a good start for Oliver's Industrial tractors. *Glenn B. Bazen/ Lull Collection*

purchase four-wheel-drive axles, engines, transmissions, hydraulic components, and then build the frame and other necessary parts to complete the vehicle.

Ware Machine Works

Ware Machine Works Inc., Ware, Massachusetts, made a front-end loader and a backhoe for the Industrial 88. John Pilch, owner of Ware, became quite interested in the new 88, its advanced design and versatility. He was partly responsible for getting the Oliver Industrial wheel tractors promoted through Oliver's Industrial sales department.

John Pilch had started his business in textile plants that were vacated when the companies moved south to the Carolinas and

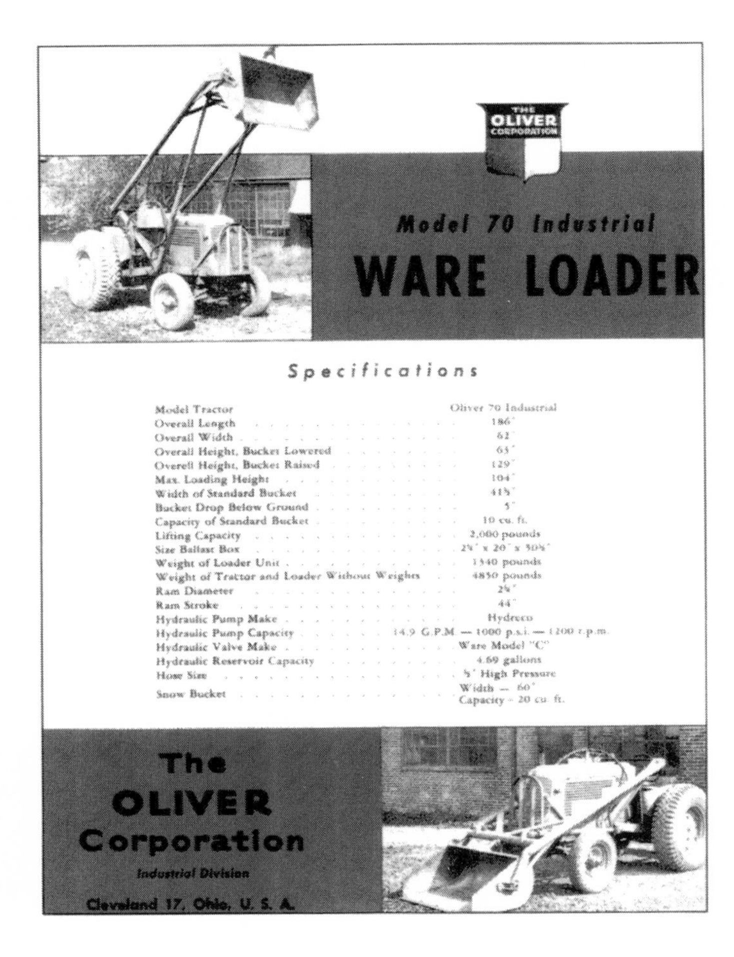

Specifications

Model Tractor	Oliver 70 Industrial
Overall Length	186"
Overall Width	62"
Overall Height, Bucket Lowered	63"
Overell Height, Bucket Raised	129"
Max. Loading Height	104"
Width of Standard Bucket	41⅝"
Bucket Drop Below Ground	5"
Capacity of Standard Bucket	10 cu. ft.
Lifting Capacity	2,000 pounds
Size Ballast Box	2⅞" x 20" x 50⅝"
Weight of Loader Unit	1340 pounds
Weight of Tractor and Loader Without Weights	4850 pounds
Ram Diameter	2⅝"
Ram Stroke	44"
Hydraulic Pump Make	Hydreco
Hydraulic Pump Capacity	14.9 G.P.M. — 1000 p.s.i. — 1200 r.p.m.
Hydraulic Valve Make	Ware Model "C"
Hydraulic Reservoir Capacity	4.69 gallons
Hose Size	⅝" High Pressure
Snow Bucket	Width — 60" Capacity – 20 cu. ft.

The Ware Loader was a popular use of the Oliver Model 70 Industrial tractor. ***Floyd County Historical Society***

Virginia for the low cost of labor and fewer union problems. My visits to his plant were very stimulating. During each visit, he would go to his patent file and pull out a patent on some new device that he had invented. We usually did some brainstorming on other possible tractor attachments that would be worthwhile to pursue.

His hospitality was great! He provided sleeping accommodations at his retreat home in the mountains nearby. Once he gave me a fishing pole and two worms, saying, "You are going to catch the fish for our supper." I went to a large pond nearby and easily caught several catfish of good eating size. It seems that John often fed his fish and they were eager to bite on most anything thrown to them including a bare hook.

One visit was during the fur trapping season. John insisted upon getting some of his friends together to go coon hunting that evening. He had hunting outfits of all sizes in his mountain home. We hunted until midnight and then had a snack before retiring. We were tired from following the dogs on the trail of a smart coon. The coon went in and out of streams of water and a small lake several times to decoy the dogs into the water. A coon can drown a dog in the water. The coon finally ended up in a large oak tree that was still full of its fall leaves. There was no way that we could get the coon out of this tree. While we were having our snack, John's oldest and wisest dog, Queeny, treed a coon within 100 yards of his mountain home. The next night, we had coon for dinner which was specially prepared by one of John's expert cooks.

Ware developed high-capacity hydraulic pumps that could be mounted on the front, driven from the engine's crankshaft, or mounted on the rear of the Industrial 88, driven by the independent power take-off.

Ware's large backhoe provided good productivity for city service installations of sewers and utilities. It became very popular for many other earth-moving applications. In fact, the Ware backhoe set a standard for many others to follow. Ware continued to manufacture backhoes and loaders exclusively for Oliver tractors.

The Parsons Co., Newton, Iowa, made some loaders and backhoes of lower capacity than the Ware units. They were adaptable to the medium-size tractors, such as Oliver's Industrial 77.

The Danuser Machine Works, Fulton, Missouri, made a line of three-point hitch-mounted scrapers, a landscaping blade, other landscaping implements and other attachments for light industrial wheel tractors. These were sold to be used with Oliver's Industrial wheel tractors.

Maine Strait-Line Loader

Maine Steel Co., Portland, Maine, had a very interesting loader designed especially to be attached to an Oliver Industrial 88. This loader was unique in that the bucket could be loaded or emptied from the front or the rear of the tractor and transported from the front to the rear. The advantage was more productivity because one could run the tractor in a straight line to load and empty without having to turn the tractor around, saving much steering and shifting of gears. Under some conditions, the Strait-Line Loader could load gravel from a pit two or three times as fast as a conventional loader.

The main disadvantage of the Strait-Line Loader was evident when it was operated on a slope or rough terrain. It was not very stable when the load was being transported from front to rear and vice versa. A small number were sold for special applications. Most of the owners were very happy with the results of great productivity. But just at the time we were getting started with this unit, George Soule, the owner, was having difficulty with the Internal Revenue Service. He had deducted the cost of development from the company's income taxes, but the Internal Revenue Service ruled that the development was productive work that increased his property and the deductions were not allowed. Soule was forced into bankruptcy, which ended another interesting adventure.

Super 55 and 552 Forklift

The versatile Super 55 and 550 became important material-handling units. The typical industrial forklift tractor was ineffective to move material in snow, soft ground or other poor traction conditions. The K-D Manufacturing Co., Inc., Waco, Texas, made forklift towers. The differential drive behind the transmission on the Oliver Super 55 and 550 tractors could be assembled to provide six speeds in reverse and two forward speeds. This was accomplished by

This Super 55 with a forklift attached on the rear was reverse driven. It was used to move materials at the Oliver plant through snow and other poor traction conditions. It was quite successful commercially. ***Carl Rabe***

placing the large differential spiral bevel gear on the other side of the spiral bevel pinion. The tractor controls and the operator's seat were assembled so that the operator faced the rear of the tractor. The K-D forklift tower was mounted next to the large driving wheels in front of the reversed operator's position. The Super 55 became quite popular for moving outside material at our plant and many other storage facilities.

The Super 55 was then improved for better stability under a maximum load and given new styling to become the Model 552. Oliver 552 forklifts were popular for use in outside storage near factories and businesses. Because of the large tires, they were particularly useful in the winter for moving outside material in the snow. The weight of the tower and the load provided good traction in snow, ice, and on soft ground.

Davis Manufacturing Inc., Wichita, Kansas, made smaller loaders and backhoes to fit smaller tractors such as the Oliver Super 55 and 550 models.

New styling and better stability when lifting maximum loads made the 552 an improvement over the already successful Super 55 forklift.
Floyd County Historical Society

Lull Street Sweeper

During the late 1950s, Lull Engineering adapted an Oliver 550 tractor as a road and street sweeper, the Model SP-2A Road and Street Broom. The 550 tractor's operator position and controls were reversed. The engine then had no axle or support under it. The broom and the frame were behind the drive wheels and the tractor was driven backward. The spiral bevel gear could be positioned on the other side of the spiral bevel pinion to provide six speeds in reverse and two forward. The position of the engine provided good fore and aft stability.

An interesting problem developed when Lull Engineering decided to add water tanks to the tractor. These tanks provided for sprinkling water to overcome dusty conditions. The extra 3,000 pounds on the rear axles caused some axle failures. The adding of the water tanks had not been reviewed in advance with Oliver Engineering.

The Lull Road and Street Broom was a good use of the Oliver 550. The versatility of the 550 provided an ideal reverse-driven Power-Pak for sweeper applications. ***T. Herbert Morrell collection***

The axle failures led Lull to believe that Oliver did not have the strongest axle construction. We suggested that the immediate solution would be to put trusses under the axles and rear tractor, but Lull used Oliver's rear axle drawing and made some axles according to his theory. His axles failed so quickly that he came back to us for assistance in the design of the trusses. The trusses were satisfactory and to my knowledge, Lull experienced no serious failures.

990 and 995 Scrapers

The 990 and 995 Scrapers were additional special units for earth moving. The main difference between them was that the 995 had a torque converter. The scraper attachment was manufactured by a company in Columbia, South Carolina. This was a project by Oliver's tractor plant in South Bend, Indiana. Several of these units were sold to a road construction company who built side boards for the scraper. The side boards greatly increased the capacity of the scraper, but these scraper units were abused. The construction company went bankrupt, so the units were repossessed, rebuilt, and sold as used equipment.

The 990 and 995 provided ideal power for these small scrapers with a capacity of five to seven cubic yards. They became an important part of required industrial equipment in the 1960s. ***T. Herbert Morrell collection***

Hancock Scraper

The Hancock Scraper Co. in Lubbock, Texas, tested the 1900 tractor without the front wheels as a propelling unit for a scraper. During early tests I received a call from their chief engineer regarding a problem. Some of the capscrews used to attach the axle carrier to the rear frame were loose. He asked, "Could someone from your engineering department come to Lubbock and observe the problem?" We were in the process of arranging transportation when he called back and said, "Forget my earlier telephone call. I have just returned from the test site. The test unit fell about 24 feet down an almost vertical embankment and remained intact. I am no longer concerned about the strength of your unit."

Massey-Ferguson 98 Tractor

During the late 1950s, Massey-Ferguson was looking for some large Wheatland tractors to sell until they could design and introduce their own large tractor. They came to Oliver, so we built a tractor with the chassis of the 950, a four-cylinder Perkins diesel engine. The sheet metal, decals, and color were all as specified by Massey-

Massey-Ferguson contracted with Oliver to make a large tractor for them until they were ready to produce their own. The Oliver 990 was the right size and power to meet the company's general specifications, so it was used to produce the Massey-Ferguson 98 standard tractor. *Carl Rabe*

Ferguson. The front frame was new to adapt the Perkins engine. Several hundred were manufactured before we introduced the 1800 and 1900, and before Massey-Ferguson started to manufacture large tractors.

David Brown

The market for small tractors changed in the early 1960s. Professional people were buying small farms or large acreages for building homes. They worked at their regular job during the week, and in the evenings and on weekends they farmed their hobby farm. Small foreign tractors were appealing because of the size and lower cost. Roy Randt, manager of product planning for Oliver, negotiated with David Brown of England to sell their Model 50 as the Oliver 500 with Oliver colors, decals and grills.

The Oliver 1250 was made by Fiat in Italy. It had a certified PTO power of 38.5 horsepower when operated at a maximum speed of 2,500 rpm. These tractors were overdesigned, which gave them an extra long lifetime.
T. Herbert Morrell collection

Fiat

Oliver's contract with David Brown for Oliver 500 tractors was canceled when David Brown started selling their tractors in the United States. In 1964, J. D. Wormley, executive vice president of Oliver, started working with Fiat of Italy to establish a new source for small tractors. When I became coordinator of outside products in 1965, the Fiat tractors became my responsibility. The first gasoline 1250 tractors were not satisfactory, so Oliver made a number of suggestions. Fiat redesigned the small tractors, accepting Oliver's suggestions for the new diesel design.

In 1966, the larger Fiat tractor became Oliver's 1450 model. Both the 1250 and 1450 tractors were sent from the plant in Modena, Italy, to Genoa and then by ship to Jacksonville, Florida. Oliver then shipped them to Decatur, Georgia, where American tires and lights were installed, and Oliver decals and grills were added. It was

Oliver 1365s were made by Fiat between 1971 and 1975. This 1365 was still in use in 2010, nearly 40 years later when this photo was taken. On the tractor is four-year-old Oliver Carter Lumb, who was named after the tractors because he is the great-grandson of T. Herbert Morrell. *T. Herbert Morrell collection*

a pleasure to meet Dr. Carneo, president of Fiat, when I was in Italy in November 1969. He helped to expedite getting repair parts for the Oliver 1250 and 1450 tractors. Fiat became a very good source for small Oliver tractors.

REFERENCES

R. B. Gray. *The Agricultural Tractor: 1855-1950.* American Society of Agricultural Engineers: St. Joseph, Mich.

G. R. Gregg. *Progress in Tractor Power from 1898.* Brochure from Charles City plant, White Farm Equipment (1976) Reprinted by Alan King.

Hart-Parr Catalog #13. (1912) Reprinted by Alan King.

John E. Janssen. *Hart-Parr Tractor's Contribution To The Advancement of Agriculture.*

Alan King. *Oliver Hart-Parr 1898-1975.* Data Book No. 4 (1990).

Alan King. *An Advertising History 1929 to 1940.* (1981).

Oliver Corporation, Progress in Tractor Power from 1898. (1965) Reprinted by Alan King.

P. A. Letourneau. *Oliver Tractors: Photo Archive.* Iconografix: Minneapolis, Minn. (1993).

Robert N. Pripps and Andrew Morland. *Farm Tractor Color History: Oliver Tractors.* Motorbooks International, Osceola, WI. (1994).

Sherry Schaefer, ed. *The Hart-Parr Oliver Collector.* The Hart-Parr Oliver Collectors Association, P.O. Box 685, Charles City, Iowa 50616 (Published quarterly).

Wendell M. Van Syoc, ed. *An Historical Perspective of Farm Machinery*. Society of Automotive Engineers, Warrendale, PA. (1980).

C. H. Wendel. *Oliver Hart-Parr.* Motorbooks International, Osceola, WI. (1993).

White Farm Equipment. *Field Boss. Form R.1673* (1975).

APPENDIX A

Tractor Models

Hart-Parr Models

Model	Comment	Years Manufactured
17 - 30*	Nos. 1, 2	1901 - 1906
18 - 30	No. 3	1903
22 - 45		1903 - 1906
30 - 60	Old Reliable (On display at Charles City)	1907 - 1918
40 - 80		1908 - 1914
15 - 30		1910 - 1912
60 - 100	9-ft. diameter drive wheels	1911 - 1912
20 - 40		1912 - 1914
12 - 27		1914 - 1915
15 - 22	Little Red Devil - single wear drive wheel	1914 - 1916
18 - 35		1915 - 1918
12 - 25	First low-silhouette tractor, water-cooled.	1918
15 - 30A		1918 - 1922
35	Road King - extra lugs for pulling road graders.	1919
10 - 20B		1921 - 1922
10 - 20C		1922 - 1924
15 - 30C		1922 - 1924
22 - 40		1923 - 1927
16 - 30E		1924 - 1925
16 - 30F		1926

Model	Comment	Years Manufactured
12 - 24E		1924 - 1928
18 - 36G		1926 - 1927
18 - 36H		1927 - 1930
28 - 50		1927 - 1930
12 - 24H		1928 - 1930

*** In these model numbers the first number indicates the rated draw bar horsepower and the second is the rated belt horsepower.**

Oliver - Hart-Parr Models

Model	Comment	Years Manufactured
18 - 28		1930 - 1937
18	Industrial (18 - 28)	1931
80	Industrial (Improved 18 - 28)	1932 - 1947
18 - 27	Row Crop - Single Front Wheel	1930
18 - 27	Row Crop - Dual Front Wheel	1931 - 1937
28 - 44		1930 - 1937
99	Industrial (28 - 44)	1932 - 1947

Oliver Models

Model	Comment	Years Manufactured
90 & 99	Standard	1937 - 1952
35	Industrial (90 Standard)	1937 - 1938
50	Industrial (99 Standard)	1937 - 1948
70	Row Crop	1935 - 1948
70	Standard - Style #1	1936 - 1937
70	Standard - Style #2	1938 - 1948
70	Orchard	1937 - 1948
25	Airport (70 Standard)	1937 - 1948
80	Standard	1937 - 1948
80	Row Crop	1938 - 1948
EOA	Engine Over Transmission - for Road Graders	1938 - 1967
60	Row Crop	1940 - 1948
60	Standard	1942 - 1948

Model	Comment	Years Manufactured
60	Industrial	1946 - 1948
900	Industrial	1946 - 1951
66	Fleetline - Row Crop	1949 - 1954
66	Fleetline - Standard and Industrial	1949 - 1954
77	Fleetline - Row Crop	1949 - 1954
77	Fleetline - Standard and Industrial	1949 - 1954
88	Fleetline - Row Crop	1949 - 1954
88	Fleetline - Standard and Industrial	1949 - 1954
99	Six-Cylinder	1953 - 1957
Super 55		1954 - 1958
Super 66		1954 - 1958
Super 77		1954 - 1958
Super 88		1954 - 1958
Super 44		1957 - 1958
Super 99		1957 - 1958
550		1958 - 1975
660		1959 - 1964
770		1958 - 1965
880		1958 - 1963
950		1958 - 1961
990		1958 - 1961
995		1958 - 1961
990 Scraper		1955 - 1957
995 Scraper		1958
440		1960
500	Made by David Brown in England	1960 - 1963
550 & 551	Forklift	1960 - 1964
552	Forklift	1964 - 1975
1800A		1960 - 1962
1800B		1962 - 1963
1800C		1963 - 1964
1900A		1960 - 1962
1900B		1962 - 1963
1900C		1963 - 1964
1600		1962 - 1964
1250	Made by Fiat in Italy	1965 - 1969

Model	Comment	Years Manufactured
1450	Made by Fiat in Italy	1966
1650		1964 - 1969
1750		1964 - 1969
1850		1964 - 1969
1950		1964 - 1967
1550		1965 - 1969
1255	Made by Fiat in Italy	1969 - 1971
1265	Made by Fiat in Italy	1971 - 1975
1355	Made by Fiat in Italy	1969 - 1975
1365	Made by Fiat in Italy	1969 - 1975
1465	Made by Fiat in Italy	1973 - 1975
2050		1968 - 1969
2150		1968 - 1969
1555		1969 - 1975
1655		1969 - 1975
1755		1970 - 1975
1855		1973 - 1974
1955		1969 - 1975
1865	G 950	1971
1870	G 955 - Made for Cockshutt	1973 - 1974
2270	G 1355 - Made for Cockshutt	1972 - 1974
2055	MMG 1050 LP Gas and Diesel	1971
2155	MMG 1350 LP Gas and Diesel	1971
2255		1972 - 1975
2655		1971 - 1972
244 - 80	Industrial	1973 - 1975

White Models

Model	Comment	Years Manufactured
2 - 50		1976 - 1980
2 - 60		1976 - 1979
700		1976 - 1980
2 - 70		1976 - 1982
2 - 85		1976 - 1982
2 - 88		1982 - 1987

Model	Comment	Years Manufactured
2 - 105		1974 - 1982
2 - 110		1982 - 1987
2 - 135		1976 - 1987
2 - 150		1975 - 1976
2 - 155		1976 - 1986
4 - 150	Articulated	1974 - 1978
4 - 175	Articulated	1979 - 1982
2 - 180		1977 - 1986
4 - 180	Articulated	1975 - 1978
4 - 210	Articulated	1978 - 1982
4 - 225	Articulated	1983 - 1987
4 - 270	Articulated	1983 - 1988
2 - 30	2- and 4-Wheel Drive	1978 - 1984
2 - 32	2- and 4-Wheel Drive	1979 - 1986
2 - 35	2- and 4-Wheel Drive	1979 - 1984
2 - 45	2- and 4-Wheel Drive	1979 - 1981
2 - 62	2- and 4-Wheel Drive	1979 - 1981
2 - 55	2- and 4-Wheel Drive	1982 - 1987
2 - 65	2- and 4-Wheel Drive	1982 - 1987
2 - 65	High Clearance 4-wheel drive	1986 - 1987
2 - 75	2- and 4-Wheel Drive	1982 - 1987
FB 16	2- and 4-Wheel Drive	1986 - 1989
FB 21	2- and 4-Wheel Drive	1986 - 1989
FB 31	2- and 4-Wheel Drive	1986 - 1989
FB 37	2- and 4-Wheel Drive	1986 - 1989
FB 43	2- and 4-Wheel Drive	1986 - 1989
60 & 80		1989 - 1991
100		1987 - 1989
120		1987 - 1989
125		1990 - 1991
140		1987 - 1989
145		1991
160		1987 - 1989
170		1990 - 1991
185		1986 - 1989
195		1990 - 1991

APPENDIX B

Safety Standards

One of the most important parts of developing a new line of tractors is to know and understand technical standards and recommendations that apply to tractors. Standards and recommendations can be very effective safety devices to reducing injuries and accidents on the farm and in other workplaces.

Standards are also important to allow interchangeability between tractors and implements made by different manufacturers. During the late 1920s, there were approximately 2,500 special hook-up packages required to fit all power take-off-driven implements to all makes of tractors. Standards and recommendations were established so that by the early 1940s, any implement designed to the standards could be hooked up to any tractor designed to the standards without the need for a special hook-up package.

Standardizing Organizations

The standardizing organizations for farm and light industrial equipment consist of several technical societies. The Society of Automotive Engineers (SAE) is responsible for standards and other documents involving tractors and components. Other aspects of agriculture and agricultural machinery are covered by the American Society of Agricultural Engineers (ASAE). There are some standards, such as those on the power take-off, which affect both agricultural tractors and implements. Standards in these overlapping areas are a cooperative effort between ASAE and SAE.

Members of ASAE and SAE are individual engineers and associates, so they could not commit their employers for funds and time to develop in-depth proposals. The companies belonged to the Farm Equipment Institute (FEI), which was started in 1895 as a manufacturer's trade association. It was later superseded by the Farm and Industrial Equipment Institute (FIEI). FIEI committees were set up to research and propose standards to be published by ASAE and SAE. It was very effective to support and assist in the development of standards because through FIEI, companies could establish engineering projects to cover the funds and the time spent by their employees in the development of standards. Oliver projects contributed as much as $50,000 per year.

Standards issues were considered by the FIEI Advisory Engineering Committee, which generally consists of the chief tractor engineer and a chief engineer from an implement plant from each farm equipment company. When standards were developed or changed by the Advisory Engineering Committee of FIEI, they were referred to both ASAE and SAE for approval. When approved by ASAE and SAE, standards are published in the *ASAE Year Book* and *SAE Handbook*. Those standards of worldwide significance are also submitted to American National Standards Institute (ANSI) to be considered by International Standards Organization (ISO).

These organizations have no power to enforce standards, so compliance by equipment manufacturers is voluntary. However, adherence to published SAE and ASAE standards are often a determining factor in product liability lawsuits, so voluntary compliance is in the best interest of the manufacturers.

I was deeply involved with the standards activity of the FEI and later the FIEI, representing Oliver from 1951 to 1970, and Owatonna Manufacturing Co. from 1970 to 1977. I also served on numerous committees of the SAE, the ASAE, and the American National Standards Institute. FIEI has now been superseded by the Equipment Manufacturer's Institute. Standards and recommendations continue to be developed by committees active today.

California Safety Orders

During the mid 1950s, a long list of safety items was sent from the state of California to FIEI for consideration. It covered such items as step ladders. The Advisory Engineering Committee was assigned the items specifically for tractors and implements. Among the items were requests to have all tractors with the same gear shift pattern and controls all located in the same position and actuated in the same direction. After the meeting, Bill Coultas of John Deere said, "The next thing they will want to standardize the number of engine cylinders." Someone asked, "How many cylinders would you recommend?" Bill replied, "Any number as long as it is no more than two." Someone then asked, "How about that four-cylinder gasoline engine that you use to start your two-cylinder diesel?" Bill's reply was, "You would have to bring up that thing."

Drawbar and Hitch Standards

One of the first known standards for tractors was the height of the top of the drawbar to the ground. The standard was originally established in 1917 and was revised in 1937. The revised standard was 12–15 inches for tractors with 0–50 drawbar horsepower, 14–18 inches for tractors with 51–125 drawbar horsepower, and 16–21 inches for tractors with 126 to 175 drawbar horsepower. The revised standard was jointly adopted by the American Society of Agricultural Engineers (ASAE) and the Society of Automotive Engineers (SAE) in 1937 to be published in their handbooks.

Approximately 40 years later the three-point hitch standard was approved. This was necessary because of the various sizes of implements. The standard recommends sizes of hitch points so that the hitch and implement have compatible dimensions and enough strength. Besides this interchangeability issue, this standard also addresses a safety issue. Use of the wrong size hitch for a particular implement could create a dangerous situation.

Power Take-Off Standards

More effort went into creating power take-off (PTO) standards than any other standards. The first PTO standard was established based on a report to the SAE in 1923. The standard consisted of 536 rpm and clockwise rotation when viewed from the rear of the tractor. Wayne Worthington, director of engineering of John Deere, was involved in creating this standard. The ASAE became very active in these standards soon after 1924. Splined shafts, dimensions, and other necessary information were added later. W. Leland Zink,

General Implement Co. was very active in ASAE and was chairman of the ASAE Power Take-Off Committee during the late 1920s. His committee was involved primarily with implements, implement drive lines, and tractor output drive shafts.

The 540 rpm standard PTO could not be used for implements requiring more than 50 horsepower. The shaft would become overloaded and would fail easily at power requirements above 50 horsepower. Implements and tractors with more than 50 horsepower were anticipated in the near future. The same diameter shaft of 1 3/8 inches with involute splines and driven at 1,000 rpm could be used up to 100 horsepower. For the larger tractors, a 1 3/4 inch diameter with involute splines and a capacity of 100–160 horsepower could be used.

But the 1,000-rpm PTO was not considered very effective in 1956. Oliver at Charles City was host to the FIEI subcommittees and the Advisory Engineering Committee during April, 1957. Special tractors were prepared by Oliver for the study and review of 1,000 rpm by the Power Take-Off Subcommittee and the Advisory Engineering Committee. Recommendations were made and the next year the SAE Tractor Technical Committee approved the 1,000-rpm PTO Standard J719.

The driveline between the tractor and the implement is dangerous unless the rotating shaft is well protected by shields. There was a fatal accident at a Floyd County, Iowa, home for the elderly. The tunnel-type shields had been removed and the operator stepped close to the shaft that drives the implement. His coat wrapped up on the shaft. This occurred in 1958 and at the next meeting of the subcommittee and at the Advisory Engineering Committee, I made a motion to specify the integral rotating shield as a standard. Some objected because the bearings in the rotating shield did not have satisfactory life. I then requested that the Chairman of the Subcommittee on Power Take-Off establish a task group to develop standards for bearings. Guests from the integral rotating shield manufacturers were invited to attend the next meeting of the Power Take-Off Subcommittee. We received their recommendations which resulted in greatly improved bearings. This specification of the integral rotating shield within the standard continued until recently.

Lighting and Marking of Farm Equipment

One of my first subcommittee meetings was in 1951 at International Harvester Co.'s experimental farm in Hinsdale, Illinois. Some reports had been received concerning accidents caused by automobiles approaching slow-moving farm equipment on highways at high speed. We reviewed several suggestions and chose a flashing electric light mounted on the left wheel guard which showed amber from the front and red from the rear. But that proposal was unacceptable because the Uniform Vehicle Code reserved flashing red rear lights for emergency vehicles.

The next proposal was a light mounted on the left wheel guard with 15 feet of electric wire so that if the light on the tractor was obscured from the rear, the light could be moved to the extreme left side of the implement. This light showed amber from the front and the rear. The light and other specifications from SAE Standards were used.

In 1964, Oliver hosted the FIEI Lighting and Marking Subcommittee meeting in Charles City, Iowa. The purpose was to test proposals by any of the subcommittee members. The Charles City Country Club and Wildwood Park was reserved after sundown to conduct these tests. The country club had curves, a hill, a grade, and a level road. Distances could be measured under all of these conditions. These tests also included a proposed Slow Moving Vehicle Identification Emblem (SMV), which became ASAE 276 and SAE Standard J943 and was first printed in the 1966 ASAE and SAE Handbooks as a standard.

During the daytime meetings, the subcommittee agreed on which tests to perform each night. The lighting equipment was then assembled by Oliver's Experimental Department and the subcommittee reconvened at the country club after dark. On the third night, we tested the equipment on the highway. A tractor with the SMV emblem, red tail light, and two amber flashing lights mounted on each the left and right wheel guard was driven on Highway 14 west of Charles City. We were pleased that the traffic slowed and was very cautious when approaching the tractor. After these tests were completed, Waldo Seiple of John Deere and I wrote the proposed standard for review at the next subcommittee meeting.

The only negative votes were from the members who did not attend the tests at Charles City. The standard was finally printed in the *ASAE Handbook* as S279 and SAE J137 in 1970. Some states do not use these standards, but have their own standards. *A Compendium of State Laws Relevant to Farm Equipment* prepared by FIEI in 1981 documents the differences between the state laws.

This SMV emblem became generally accepted. However, the Amish use the same SMV emblem in black and white because it is against their religion to use the bright red and orange colors.

Safety Cartoons

Oliver was a leader in the development of technical publications for agricultural equipment. The *Oliver Operator's Manual, Parts List,* and *Service Manual* were used as models when preparing ASAE Engineering Practice EP363 and SAE J1035 Recommended Practice on Technical Publications for Agricultural Equipment. The *Oliver Super 55 Operator's Manual* included safety cartoons drawn by Bill Phillips, an illustrator in the service department in Charles City.

NEVER DRIVE TOO CLOSE TO THE EDGE OF A DITCH OR GULLY

REDUCE SPEED BEFORE TURNING OR USING ONE WHEEL BRAKE

NEVER OPERATE THE ENGINE IN A CLOSED GARAGE OR SHED

BE SURE GEAR SHIFT IS IN NEUTRAL BEFORE STARTING ENGINE

APPENDIX C

Development of Oliver's New Gasoline Engine

T. H. Morrell and K. S. Minard, Oliver Corp.

Presented at the Society of Automotive Engineers, Heavy Duty Vehicle Meeting, Milwaukee, Wisconsin, September 11-14, 1961

Reprinted with Permission from SAE Paper No. 610385 September, 1961

Society of Automotive Engineers Inc.

Abstract:

After completion of the XO-121 development (12:1 compression ratio for a tractor engine) and additional studies by Ethyl Corp., development was continued on Oliver's family of gasoline engines. An important objective was to improve power and economy through better utilization of fuel qualities by finding optimum design combinations for current and projected fuels.

Using the XO-121 combination chamber, a modified Super 88 engine was tested at various compression ratios and the results compared to those obtained for the XO-121. It became evident that there were many areas requiring further development, including combustion chamber configuration, spark plug location, manifolding, camshaft, and valve timing.

A series of tests was conducted, taking one item at a time. Evaluation and comparison of the results of these tests constituted a well-integrated, long-range development program. Results of this long-range program were very rewarding, as evidenced by the Nebraska test results of the Oliver 1800 tractor with this new gasoline engine.

Development of Oliver's New Gasoline Engine

Agriculture continues to progress each year through new techniques and improved mechanization. The farm tractor industry is contributing much to this advancement by providing more efficient engines. In part, such engines have been made possible by the continuing improvement in the quality of petroleum products. The ultimate objective, of course, is to provide the farmer with the lower overall equipment costs.

In the interest of providing our customers with the most efficient engine for best utilization of available fuels, Oliver has carried out an active engine research program since the early 1900s. The introduction in 1935 of the Oliver 70 tractor, with its high-compression six-cylinder engine, represented a milestone for utilization of fuel qualities then available. Further research continued, and in 1954 Oliver announced the results of tests on the XO-121 engine through the ASAE paper, "Looking Ahead of Tomorrow in Tractor Engine Design" by T. H. Morrell and H. K. Dommel.

The XO-121 was an experimental four-cylinder overhead-valve engine, with a bore of 3-3/4 inch and stroke of 4-1/2 inch providing a 199 cubic-inch displacement. This engine derived its name from its 12:1 compression ratio. At this ratio, experimental gasoline with octane rating considerably higher than that commercially available was required. Designed as a research engine only, the XO-121 illustrated the performance improvements which could be achieved through use of improved fuels permitting higher compression ratios. The performance of this engine was exceptional and fuel economy was especially gratifying. An observed fuel consumption of 0.285 lb/hp/hour was obtained at the flywheel with accessories. When installed in a tractor, fuel consumption of 0.400 lb/hp/hour was recorded on belt pulley power tests.

The paper by H. T. Mueller and R. E. Gish of Ethyl Corp. presented at the September 1954 SAE meeting and titled "Tractor Engine Design Requirements for Best Fuel Utilization" contained the results of an additional study conducted by Ethyl Corp. of the

XO-121 at compression ratios of 7.0 and 9.5:1. The purpose of this investigation was to relate the performance of the XO-121 engine at lower ratios and lower fuel quality to that of production engines then available. These studies proved that substantial gains in performance could be obtained by applying the XO-121 principles to present-day engines.

To apply the concepts of the XO-121 to a production engine, we selected our basic Super 88 gasoline engine for further development in 1955. This was a six-cylinder, 3-3/4-inch bore by 4-inch stroke, 265 cubic-inch, valve-in-head, wet-sleeve engine with 7.0:1 compression ratio. This paper covers the steps and considerations involved in the development of Oliver's new 1800 gasoline engine from this 1955 beginning.

Two-Piece Cylinder Head

For our development work, we decided to apply a new cylinder head of two-piece construction with an XO combustion chamber to the Super 88 engine. We had been experimenting for some time with two-piece cylinder head designs and had accumulated considerable field experience with several hundred two-piece heads on the Super 88 and other models. This development was originated by Mr. G. W. Bird, our former plant manager, who is now retired.

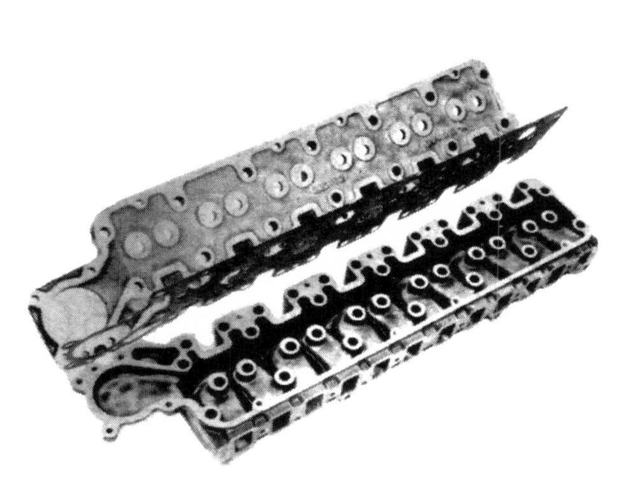

Figure 1. Two-Piece Cylinder Head

The design consists of a main lower cylinder head casting with a separate cover casting, as shown in Figure 1. The mating surfaces of the head body and cover are first machined, then assembled with a gasket between them. The remainder of the machining is accomplished as if the head was of one-piece construction. The benefits derived from such a design are:

1. Easier to cast.
2. Reduced dry sand core weight.
3. Reduced casting scrap from core shift or blow holes.
4. Easier to clean.
5. Elimination of cooling restrictions by removing all fins, core wires and core sand.
6. Easier to inspect and determine if casting is sound prior to machining.

The adaptation of the two-piece construction is desirable for more consistent cooling to permit maximum performance from a high-compression engine.

Compression Ratio Study

The first study in the development of this new engine was to evaluate the performance of the XO combustion chamber at various compression ratios and compare this performance to the yardstick previously established by the XO-121 engine studies. To accomplish this, cylinder heads were provided with compression ratios of 7.0, 8.0, 9.5, and 12.0:1. Figures 2 and 3 show cross sections of the 7:1 and 12:1 cylinder heads. Intermediate compression ratios were obtained by varying the depth of combustion chamber. The 18mm spark plugs were located on the left side of the engine between external (non-water-jacketed) exhaust ports. The valves and roof of the chamber were placed at a 12-degree angle.

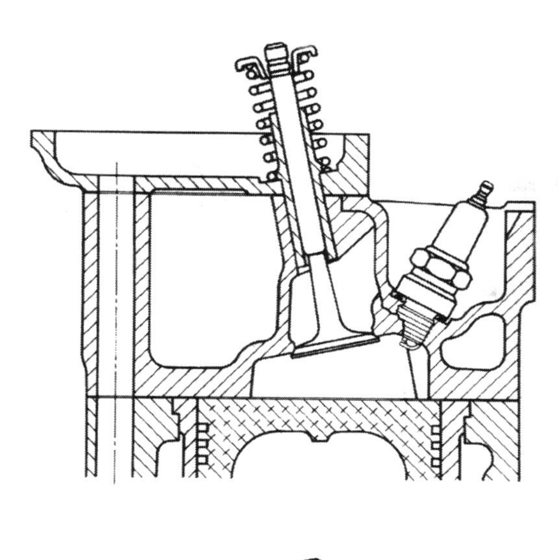

Figure 2.
7.0:1 Compression Ratio

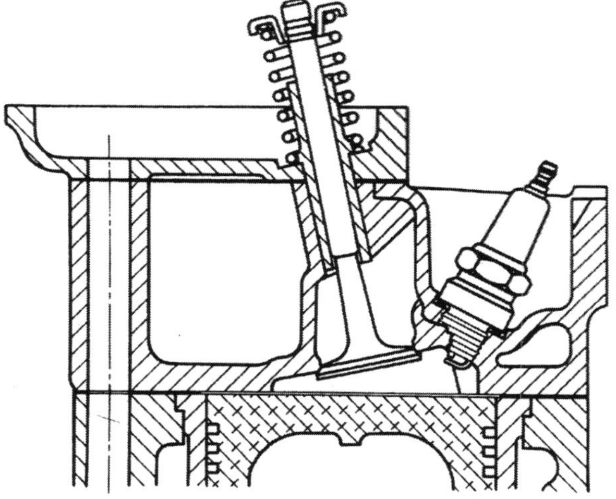

Figure 3.
12.0:1
Compression Ratio

Figure 4 shows the performance of this engine with the XO head at various compression ratios and the performance of the XO-121 at the same ratios. To eliminate the adverse effect of detonation on performance, special non-knocking high-octane fuel was used for this comparison. Although the Super 88 showed acceptable performance at the various ratios and response to changes in ratio, the results were not as good as our previously established XO-121 yardstick. This is noted by the higher brake mean effective pressure and lower specific fuel consumption of the XO-121 at each compression ratio. These results illustrate that design principles employed on one engine do

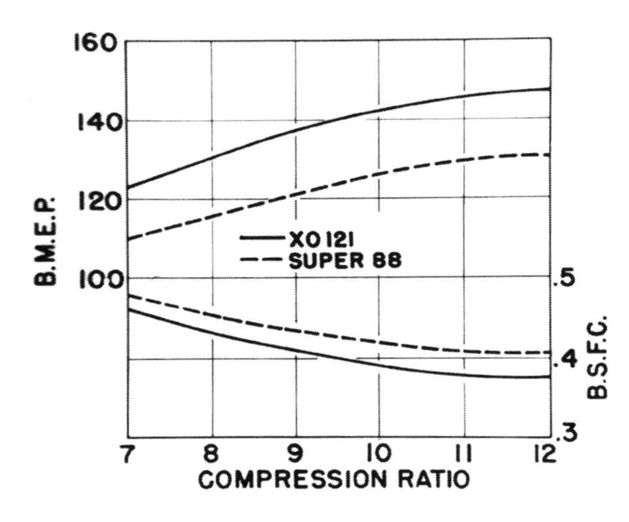

Figure 4. Performance Comparison at 1600 rpm without Fan

not necessarily hold true on another engine with a different design configuration. It became evident that additional refinement would be required to secure peak performance.

This further development of the Super 88 engine was resolved into the following phases for additional studies:

1. Structural rigidity.

2. Combustion chamber configuration.

3. Induction and exhaust system analysis.

Structural Rigidity

The results of some early tests indicated that performance might be improved by increasing structural rigidity. A prime factor in the structural rigidity studies was to provide our customer with increased engine life through great durability, in additional to improved efficiency and low cost of operation. Since we were carrying out a parallel future diesel program which might also benefit from increased structural rigidity, and considering operation at speeds considerably faster than 1600 rpm, we decided to test and evaluate a crankshaft with seven main bearings, as shown in Figure 5. At the same time, a full-pressure lubrication system with high-capacity gear pump was incorporated, replacing the metered system previously used.

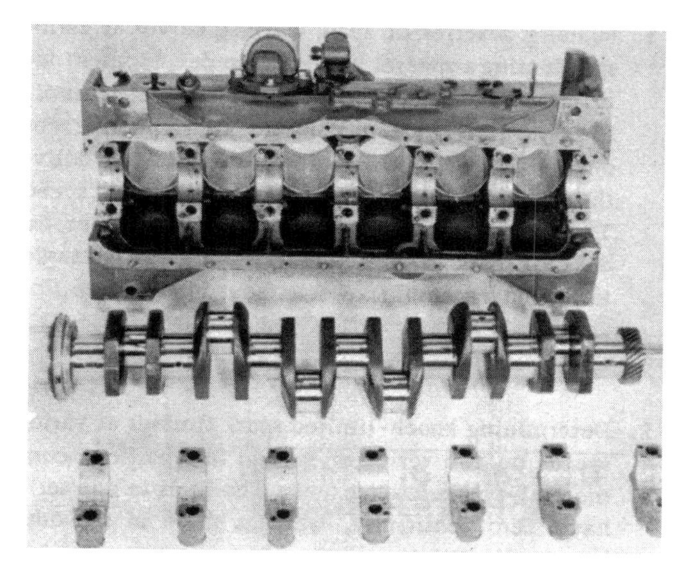

Figure 5. Crankcase with Seven Main Bearing Crankshaft

Friction studies showed no noticeable difference between the engine with seven main bearings and the production engine with four main bearings. Any added friction of the three additional main bearings could not be detected. At high speeds, power output was slightly improved and engine operation was somewhat smoother with the seven-bearing engine. Thus, we decided to adopt the seven bearing design for our overall future engine program.

Combustion Chamber Configuration

The next step was to concentrate on combustion chamber configuration and to develop a high-ratio chamber which would provide maximum power and economy using regular gasoline for this study; the 8.0:1 compression ratio head was selected.

We felt that performance could be improved by increasing the amount of turbulence in the chamber through improved squish action. There appeared to be a dead area at the shallow end of the chamber, especially at lower compression ratios. The first attempt to correct this problem was to bevel the back wall of the chamber at a 45-degree angle, as shown in Figure 6. The results of this modification were disappointing and indicated a performance loss of approximately two horsepower, apparently due to a reduction in the squish area.

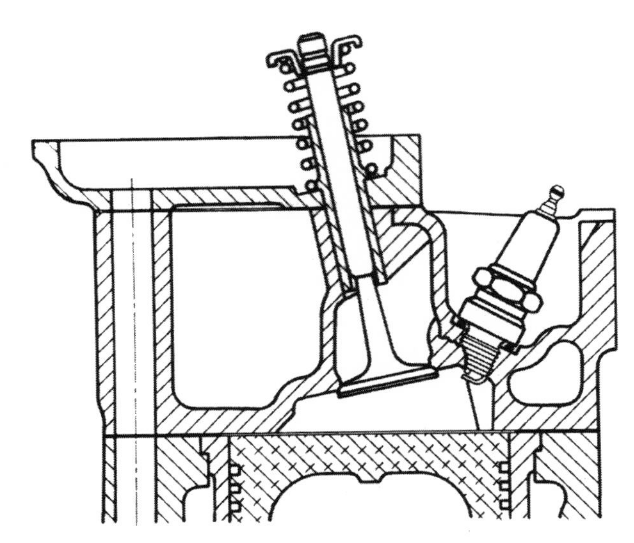

Figure 6.
Beveled Back
Combustion Chamber
8.0:1 Compression Ratio

Squish height (distance from piston to cylinder head) was also investigated, and the optimum height was determined to be 0.040 to 0.060 inch. The squish path was very important to concentrate the fuel-air mixture in the spark plug area. When spark plugs were evaluated, the projected-nose type was found to provide a slightly flatter spark traverse curve with better power at retarded spark settings. These results held true in additional developments.

The next attempt to improve turbulence was to reduce the depth of the combustion chamber by using the 9.5:1 ratio cylinder head and placing the remainder of the combustion chamber volume in the piston to obtain an overall ratio of 8.0:1. This combination is shown in Figure 7. This design, called the concave piston, provided a more compact chamber, locating the spark plug closer to the squish path.

Cooling was improved in this step, as compared to the designs shown in Figures 2, 3, and 6. The compression rings were hanged to a deeper section and moved down so that the top ring did not traverse above the water jacket. This reduced the temperatures in the ring belt, thereby providing longer ring and piston life.

Another combustion chamber configuration called the wedge or hump piston was also tested at 8.0:1 ratio. As shown in Figure 8, this consisted in machining the roof of the combustion chamber and providing a matching wedge or hump on the top of the piston.

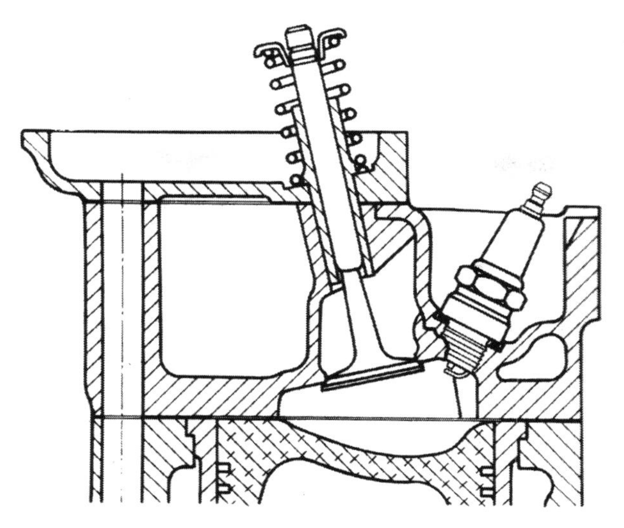

Figure 7.
Concave Piston Design

A portion of the combustion chamber volume was provided by lowering the flat deck of the piston approximately 3/16 inch below the top of the sleeve. We believed that this design, while still retaining a compact chamber, might improve performance by directing the squish directly past the valves to the spark plug and thus creating better turbulence and scavenging in the spark plug area.

To fully evaluate the performance of these chambers with fuels of different octane ratings for comparison with the flat-piston design, the knock-rating procedure previously developed by Ethyl Corp. was employed. Basically, this procedure involves the following three steps:

1. Running a series of spark traverse curves at various speeds using a special high-octane development fuel to prevent knock. The development fuel is similar to a regular-grade gasoline in all respects except octane number. The carburetor is adjusted to deliver the final calibration air-fuel ratio at each speed. The runs provide power and specific fuel consumption data for spark timings ranging from greatly retarded to beyond maximum power without interference from knock.

2. Determining knock-limited spark timings at various speeds on two series of special full-boiling, commercial-type

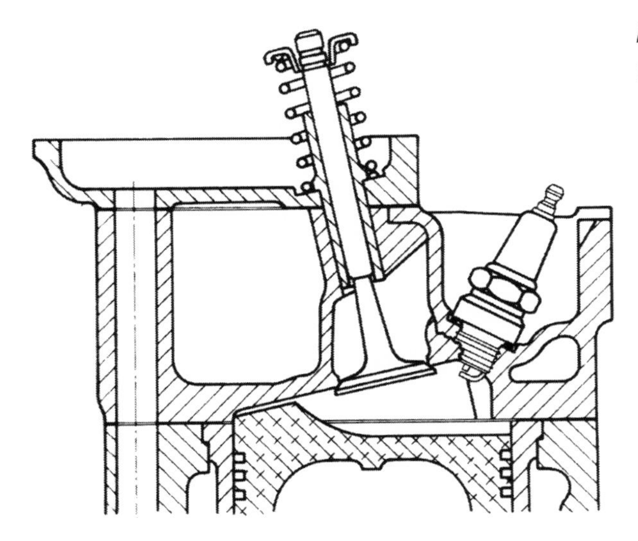

Figure 8. Wedge Piston Design

reference fuels. The fuels in one series have zero sensitivity, whereas those in the other have a sensitivity of ten. Motor octane numbers range from 72 to 87 in 5-octane increments in each series. Thus, the fuels range from 72 Motor-72 Research, to 87 Motor-97 Research. The knock-limited spark timing is that required to produce a barely audible knock (commonly called trace or borderline knock). The same air-fuel ratios used during spark traverse test on development fuel are maintained during the knock tests.

3. Interpolating between reference fuels to determine spark timings for borderline knock on fuels of any desired octane number and sensitivity, and then finding the corresponding power and specific fuel consumption data from the spark traverse curves.

For our analysis of the three combustion chambers, we selected the following three hypothetical fuels:

Low	-	80.5 Motor, 88.5 Research
Average	-	83.0 Motor, 91.5 Research
High	-	86.0 Motor, 95.0 Research

The borderline knock-limited performances of the three combustion chambers with these hypothetical fuels are shown on Figures 9, 10, and 11. In these tests, the concave piston design provided superior performance with all three fuels, indicating good fuel utilization regardless of fuel quality. Although the wedge piston did not perform well with low-quality fuel, its performance improved at a greater rate than that of the other chambers when higher quality fuels were used. Data was not taken at speeds above 2200 rpm, but the results indicated that the wedge piston might perform quite well at higher speeds. Some trouble was encountered with the wedge pistons due to a hot spot at the top of the wedge. This hot spot, which caused pre-ignition, was reduced but not entirely eliminated by redesign of the underhead of the piston.

An evaluation of the data obtained to this point indicated a possibility of reverse squish between the piston and head underneath the spark plug. This reverse squish appeared to counteract the main squish from the other side of the piston, thereby reducing the turbulence in the spark plug area. To investigate this possibility, a new cylinder head was designed, as shown in Figure 12. In this design, the valves were placed at an 11-degree angle instead of 12-degree to obtain better rocker arm action, and the combustion chamber was shifted toward the spark plug side. The spark plug remained on

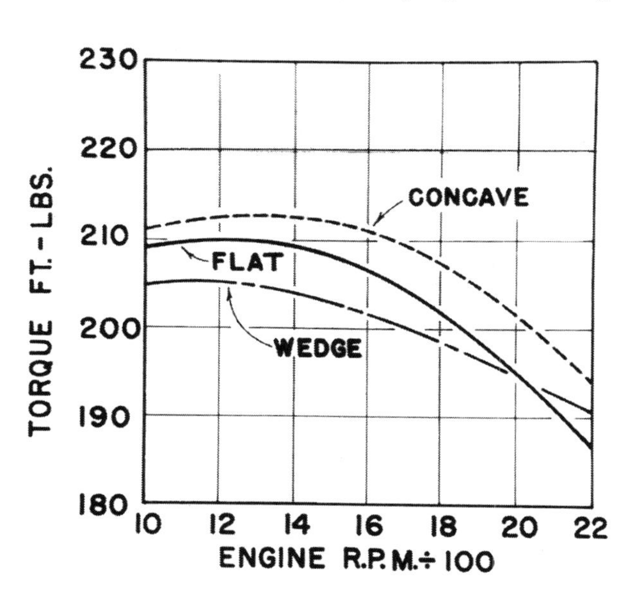

Figure 9. Borderline Knock-Limited Torque with Low Hypothetical Fuel – 8.0:1 Compression Ratio

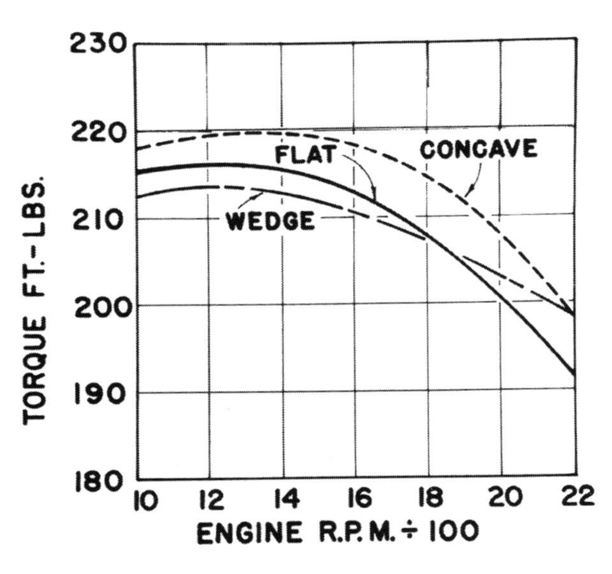

***Figure 10.* Borderline Knock-Limited Torque with Average Hypothetical Fuel – 8.0:1 Compression Ratio**

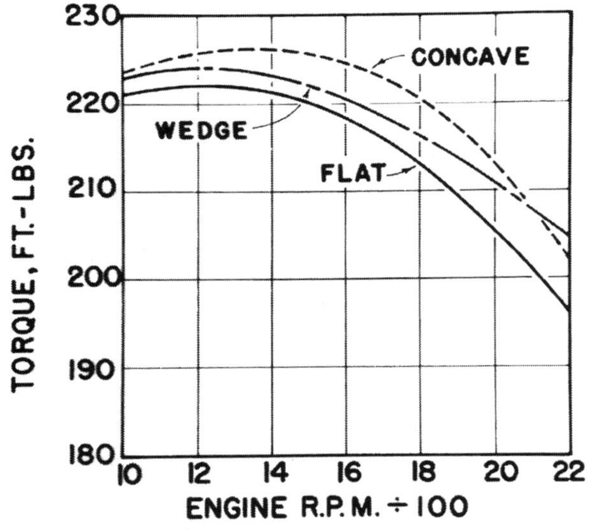

***Figure 11.* Borderline Knock-Limited Torque with High Hypothetical Fuel – 8.0:1 Compression Ratio**

the left side of the engine between exposed (non-water-jacketed) ports, but the ledge connecting the ports at the edge of the head was removed to prevent accumulation of dirt and trash around the spark plug. This head was tested with the flat piston and appeared equal to or slightly better than the previous flat-piston designs.

It became evident at various stages of the development that spark plug life could be improved materially if the plugs could be moved away from the exhaust ports. Since the distributor was on the right side of the engine, it was desirable that the plugs be on the same side. Therefore, we decided to reverse the cylinder head and combustion chamber, placing the plugs on the right side while leaving the intake and exhaust ports on the left side of the engine, as shown in Figure 13. To further improve cooling around the spark plug it was necessary to decrease the plug size from 18mm to 14mm and use a 3/4 -inch reach plug. The combustion chamber was shifted farther toward the spark plug side to eliminate reverse squish and to gain more squish area opposite the spark plug as can be seen in comparing Figures 7 and 13. Since the concave piston had continued to show improved fuel utilization over the wedge or flat-top pistons, all further development was continued with the concave design.

The name Econo-Pak was assigned to this final combustion chamber configuration. Econo-Pak was derived from the fuel economy capabilities of the compact combustion chamber.

Induction and Exhaust System Analysis

In anticipation of higher operating speeds and for better breathing, intake and exhaust valves and ports were enlarged. The reversed combustion chamber also permitted more steamlined intake and exhaust ports which reduced restriction on the intake air.

Studies of camshafts for higher speeds were made to determine if further improvements were possible. We evaluated a series of five different camshaft designs in which the lift and the opening and closing events were varied. The best camshaft design had the same timing as the original Super 88 but approximately 17 percent greater lift. The profiles of these two camshafts are compared in Figure 14. Figure 15 indicates considerable improvement in performance with the new high-lift camshaft throughout the speed range. To determine the necessary ramp design for this cam, a Lashograph was employed to determine lash loss under under all conditions of operation. This was an optical device obtained from Ethyl Corp. which visually indicates valve lash during operation.

Manifold investigations were conducted throughout the entire development program, with various designs being tested. The original exhaust manifold connected to cylinder-head exhaust ports which

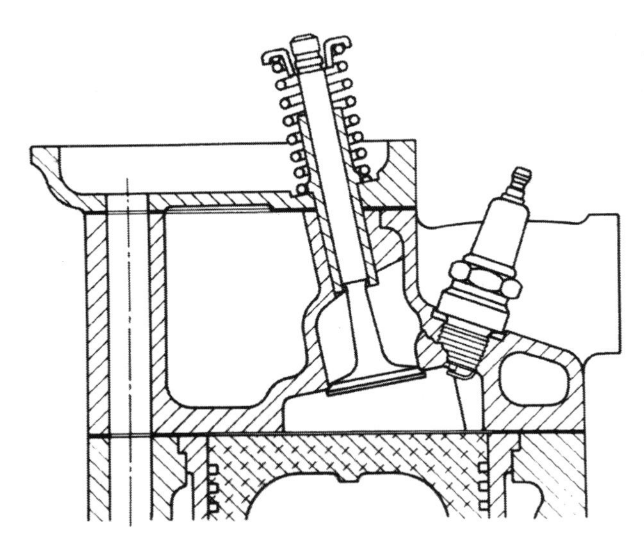

Figure 12.
Revised Head
11-degree Design

were not water-jacketed presented a warpage problem which was reduced by reversing the head and jacketing the exhaust ports. Further improvements in the exhaust manifold were made by using a special high-strength low-growth cast-iron alloy, increasing the runner size by 25 percent, and providing ribbing, as shown in Figure 16.

The intake-manifold investigations involved determination of riser and runner size, manifold shape, placement and amount of manifold heating and riser length. In the final design, heat was applied

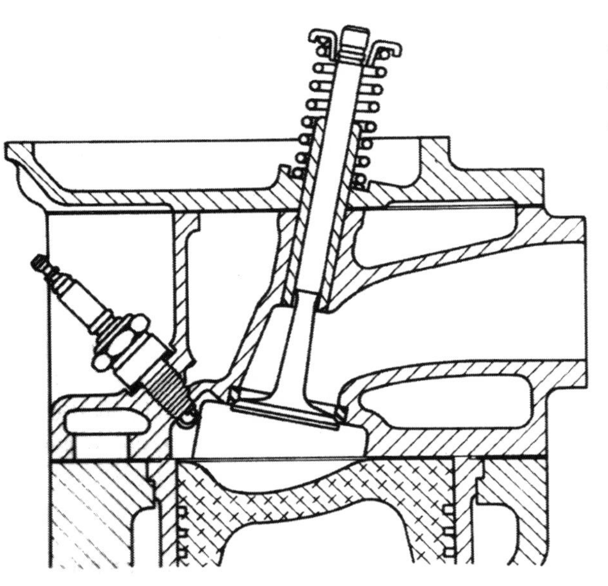

Figure 13.
Final Head and
Piston Design

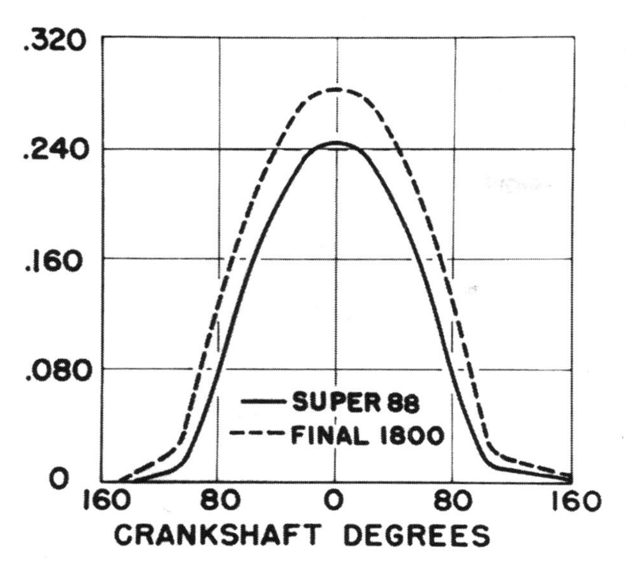

Figure 14.
Intake Cam
Profile Comparison

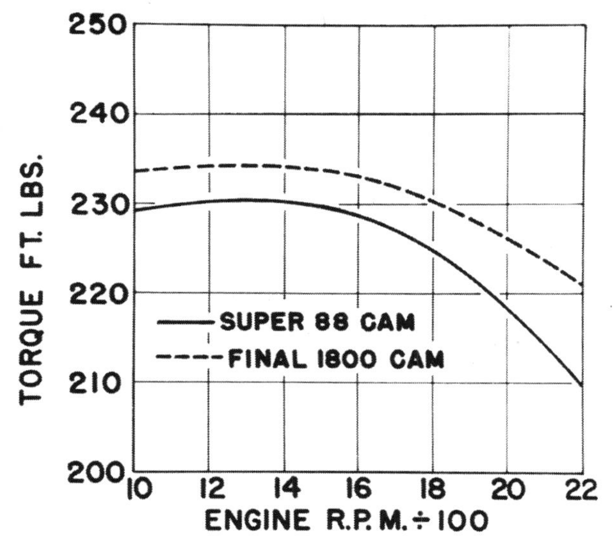

Figure 15.
Camshaft Performance
Comparison with
Non-Knocking Fuel –
14.1:1 A.F.R.

to the top of the T section above the riser, and a rectangular runner was employed to reduce any tendency for fuel to centrifuge out of the mixture. Individual porting of each cylinder was incorporated.

Final Performance

Following the completion of work on the induction system and related components, a borderline knock-rating analysis was again

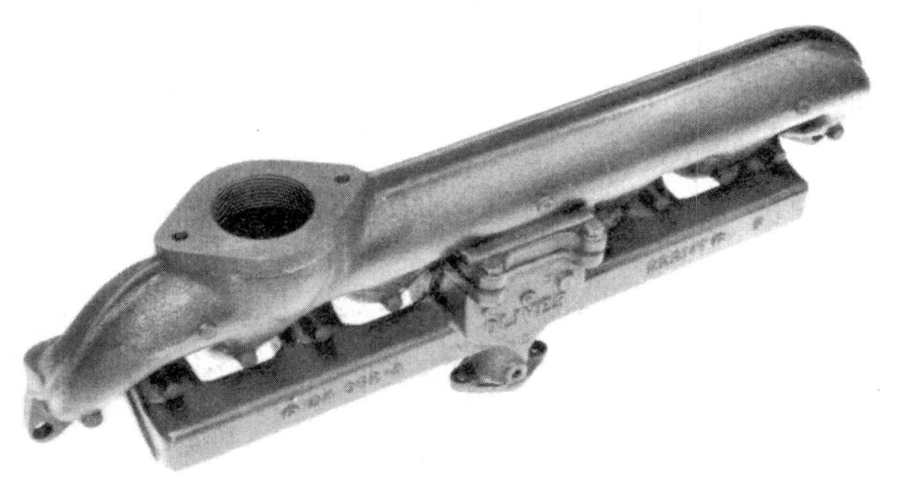

Figure 16.
Intake and Exhaust Manifold

employed to determine the maximum compression ratio at which the engine could be operated when placed in production. For this analysis, typical 1961 fuels with octane ratings as follows were selected:

Low	-	82.5 Motor, 90.5 Research
Average	-	84.0 Motor, 92.5 Research
High	-	85.5 Motor, 94.5 Research

With these fuels, it was found that the engine developed maximum knock-limited power at a compression ratio of 8.5:1. This ratio was therefore selected for production. Figure 17 shows the performance of the Oliver 1800 engine at borderline knock with these three typical fuels.

A comparison of final performance of the Oliver 1800 gasoline engine as released for production with the original XO-121 and the Super 88 with XO combustion chamber, all at 8.5:1 compression ratio, is shown in Figure 18. These curves indicate the performance improvement obtained by the additional development of this engine and show that the performance goals established by the XO-121 at the same ratio can be obtained on a production engine.

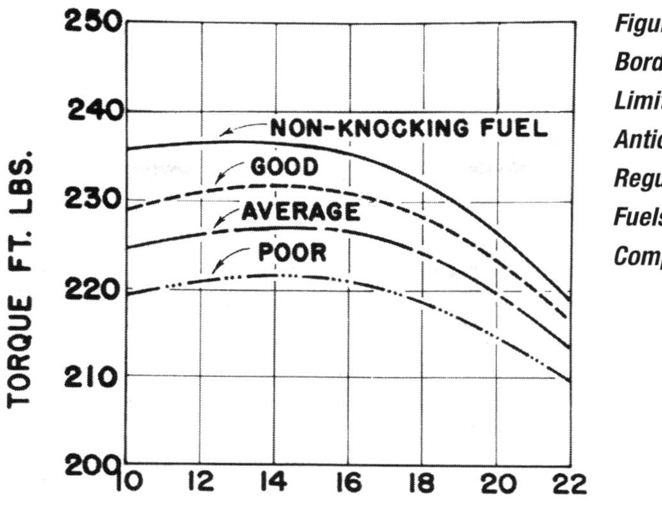

Figure 17. Borderline Knock-Limited Torque with Anticipated 1961 Regular Grade Fuels – 8.5:1 Compression Ratio

One of the main objectives of this long-range research program was to maintain good economy throughout the complete operating speed range of the engine. The results were very gratifying, as evidenced by the relatively flat fuel economy-curve in the speed range of 1000 to 2200 rpm.

This research program also permitted improvements in our diesel engines which were derived from the two-piece cylinder head

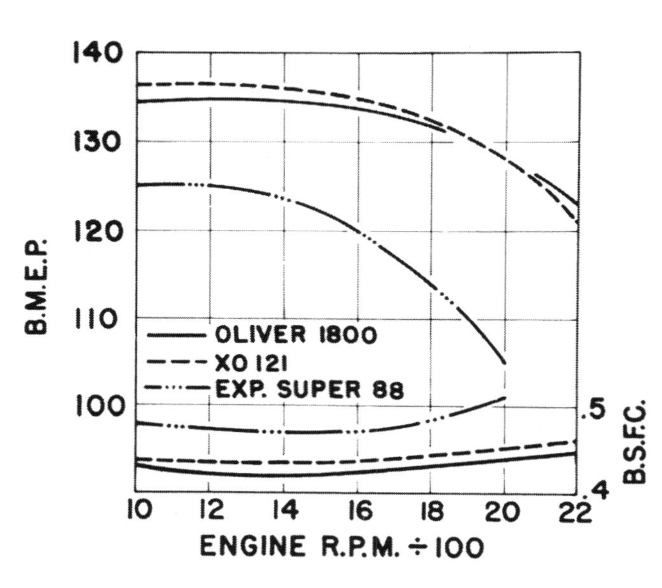

Figure 18. Final Performance Comparison with Standard Accessories, Carbureter Adjusted to 97 percent of Maximum Power

and studies of manifolding, camshaft timing and profile, valve lash, lubrication, and other design characteristics of a minor nature.

On Nebraska Test 766, the model 1800 established a new all-time fuel economy record of 0.472lb/hp-hour or 12.18 hp-hr/gal. on the maximum-power PTO test. Under the current test procedure, institute in 1959, the following fuel economies were obtained:

Test Description	Lb./hp-hr	HP-hr/gal
Max. Power – PTO	0.473*	13.18*
Varying Power – PTO		
Rated	0.481*	12.93*
½ Rated	0.651*	9.56*
Maximum	0.474	13.14*
¼ Rated	0.971*	6.41*
¾ Rated	0.527*	11.80*
Average	0.590*	10.55*
Max. Drawbar Power	0.532*	11.71*
75% Pull Drawbar Power	0.566	10.99*
50% Pull Drawbar Power	0.688	9.04*

*Denotes new record for current test procedure established by Test #766.

These records were established despite a governed speed of 2,000 rpm with power-consuming accessories such as power steering and hydraulic pumps.

In conclusion, we can say that the results of this program have been very rewarding and have enabled us to provide an efficient, reliable product for our customer.

Acknowledgements

We wish to acknowledge the excellent cooperation received from the Ethyl Corp. and Waukesha Motor Co. in assisting this project to a successful conclusion.

We thank the other suppliers for their valued aid in this program.